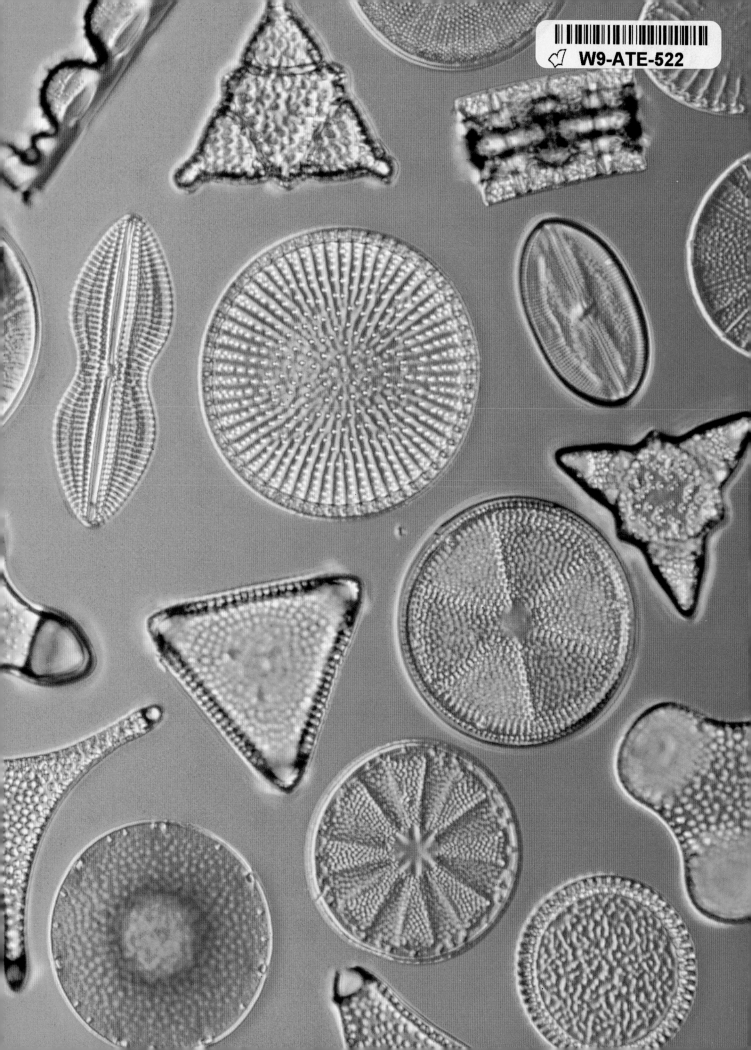

W9-ATE-522

Ohio Dominican College
1216 Sunbury Rd.
Columbus, OH 43219

INSIDE GUIDES

MICROLIFE

Written by

DAVID BURNIE

DK PUBLISHING, INC

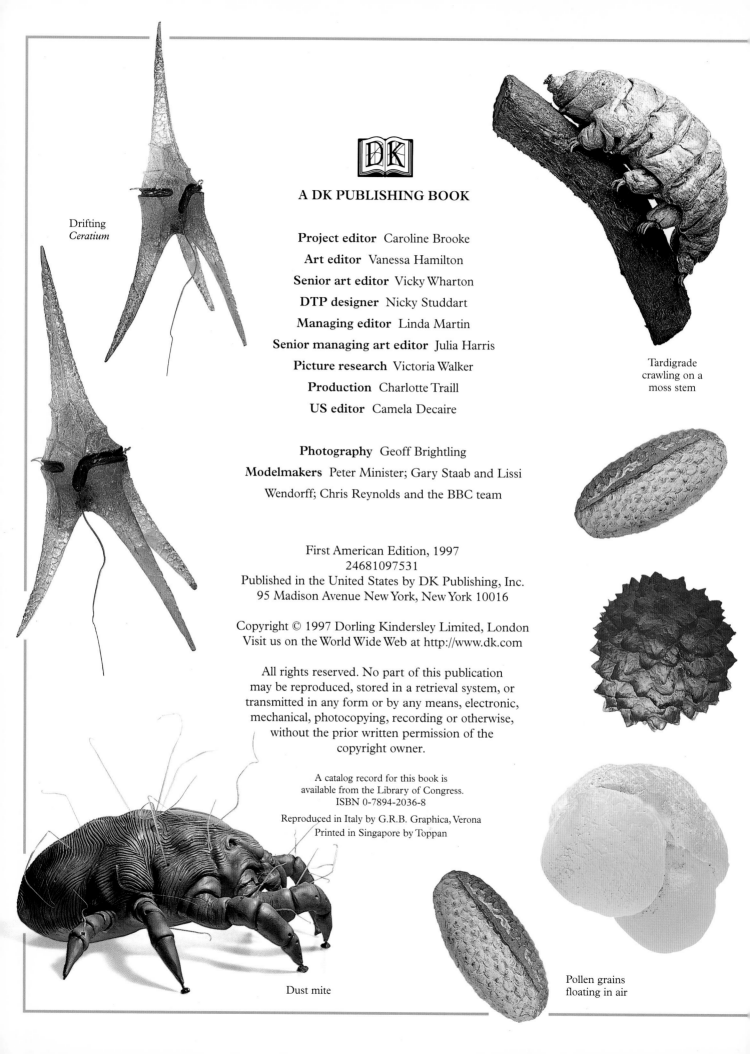

Drifting
Ceratium

Tardigrade
crawling on a
moss stem

A DK PUBLISHING BOOK

Project editor Caroline Brooke
Art editor Vanessa Hamilton
Senior art editor Vicky Wharton
DTP designer Nicky Studdart
Managing editor Linda Martin
Senior managing art editor Julia Harris
Picture research Victoria Walker
Production Charlotte Traill
US editor Camela Decaire

Photography Geoff Brightling
Modelmakers Peter Minister; Gary Staab and Lissi
Wendorff; Chris Reynolds and the BBC team

First American Edition, 1997
24681097531
Published in the United States by DK Publishing, Inc.
95 Madison Avenue New York, New York 10016

A catalog record for this book is
available from the Library of Congress.
ISBN 0-7894-2036-8

Reproduced in Italy by G.R.B. Graphica, Verona
Printed in Singapore by Toppan

Dust mite

Pollen grains
floating in air

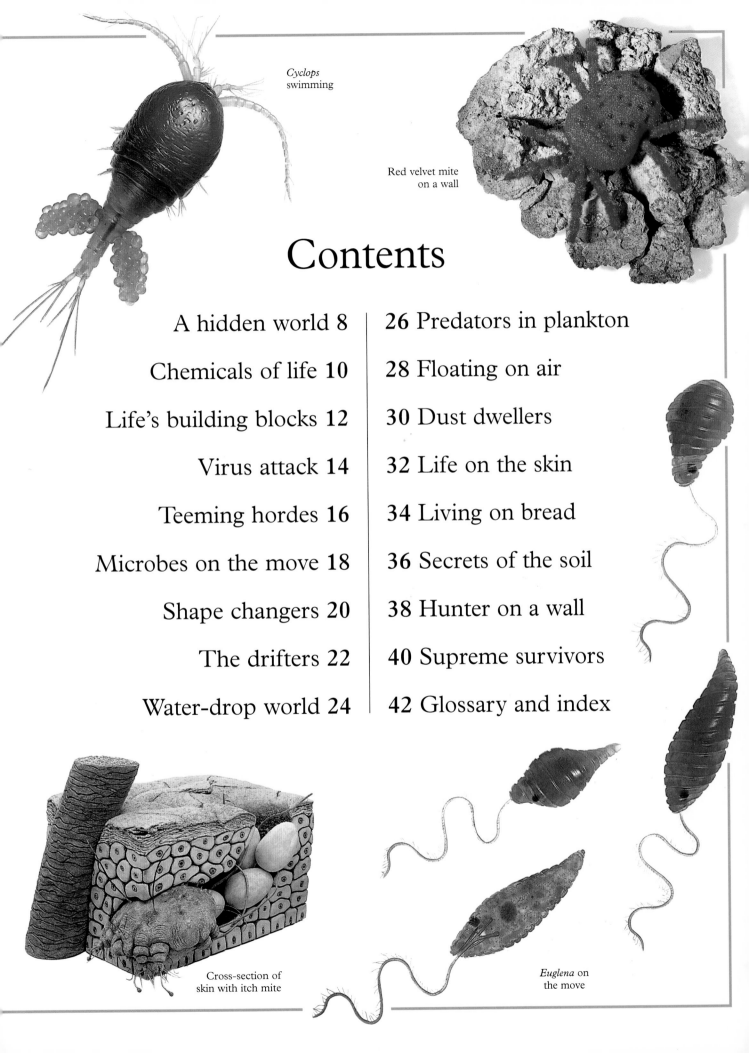

Cyclops
swimming

Red velvet mite
on a wall

Contents

Cross-section of
skin with itch mite

Euglena on
the move

A hidden world

The human eye is an amazing instrument, but one with built-in limits. It is good at seeing large things, but it has difficulty with things that are very small. When things are less than about 0.008 in (0.2 mm) across, our eyes cannot see them at all. Before the invention of the microscope in the 16th century, people had little idea of the hidden world that lay beyond the limit of sight. Now, thanks to increasingly powerful microscopes, scientists have discovered that it teems with a vast range of living things. This book explains what some of these things are, and also how they live.

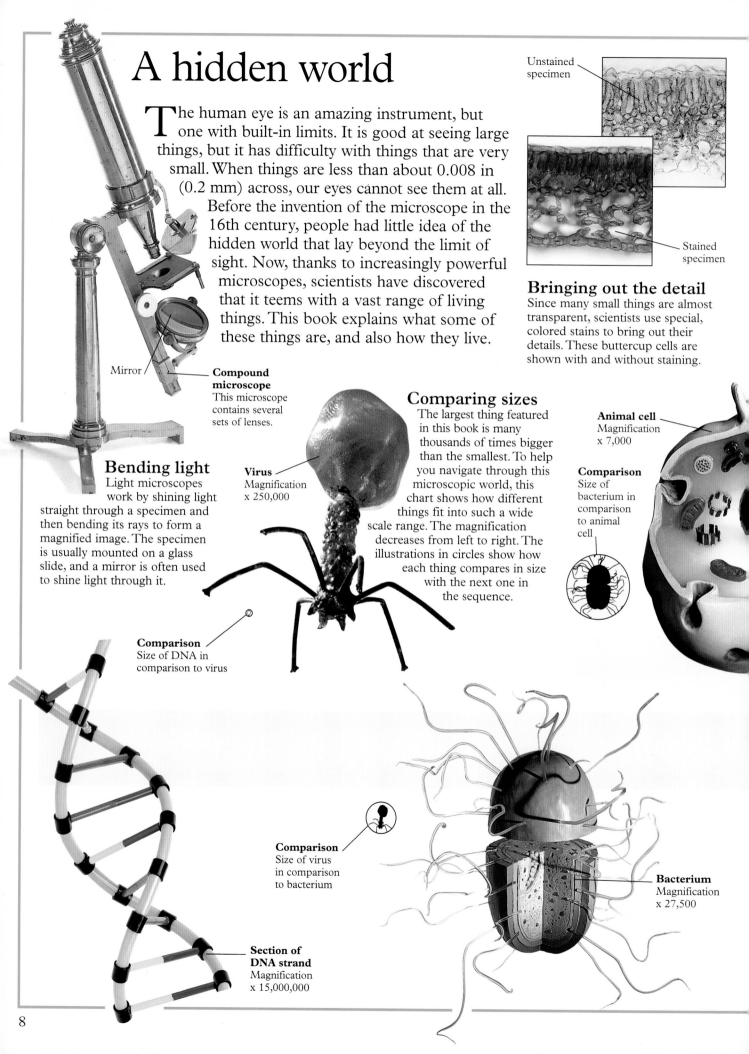

Mirror

Compound microscope
This microscope contains several sets of lenses.

Unstained specimen

Stained specimen

Bringing out the detail
Since many small things are almost transparent, scientists use special, colored stains to bring out their details. These buttercup cells are shown with and without staining.

Bending light
Light microscopes work by shining light straight through a specimen and then bending its rays to form a magnified image. The specimen is usually mounted on a glass slide, and a mirror is often used to shine light through it.

Virus
Magnification
x 250,000

Comparing sizes
The largest thing featured in this book is many thousands of times bigger than the smallest. To help you navigate through this microscopic world, this chart shows how different things fit into such a wide scale range. The magnification decreases from left to right. The illustrations in circles show how each thing compares in size with the next one in the sequence.

Animal cell
Magnification
x 7,000

Comparison
Size of bacterium in comparison to animal cell

Comparison
Size of DNA in comparison to virus

Comparison
Size of virus in comparison to bacterium

Bacterium
Magnification
x 27,500

Section of DNA strand
Magnification
x 15,000,000

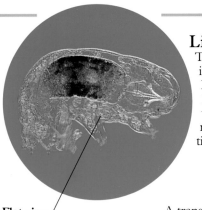

Light microscope

This tiny animal, called a tardigrade, is seen through a light microscope. Light microscopes can work with living specimens. They can magnify up to 2,000 times, but give the best results at 750 times or less.

Flat view
A typical light microscope produces a two-dimensional image.

TEM

A transmission electron microscope (TEM) works by firing a beam of electrons through a specimen. The electron beam can magnify more than 100,000 times. Unlike a light microscope, an electron microscope cannot be used with living specimens.

False coloring
Scanning electron micrographs are black and white. As with the models in this book, the colors that are added later are not always natural.

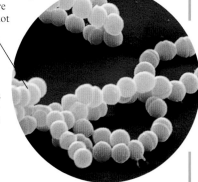

Cross-section
Electron beams are perfect for looking inside small cells.

SEM

In a scanning electron microscope (SEM), a beam of electrons bounces off a film of metal atoms that coats the specimen. This creates a three-dimensional image.

Comparison
Size of *Vorticella* in comparison to the tardigrade

Tardigrade
Magnification x 160

Visible to the naked eye

Visible under a light microscope

Visible under an electron microscope

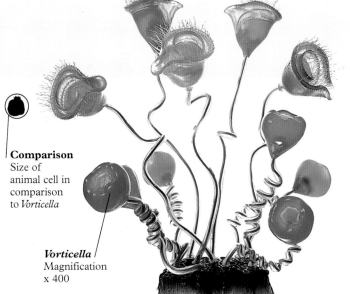

Comparison
Size of animal cell in comparison to *Vorticella*

Vorticella
Magnification x 400

Comparison
Size of tardigrade in comparison to *Cyclops*

Cyclops
Magnification x 35

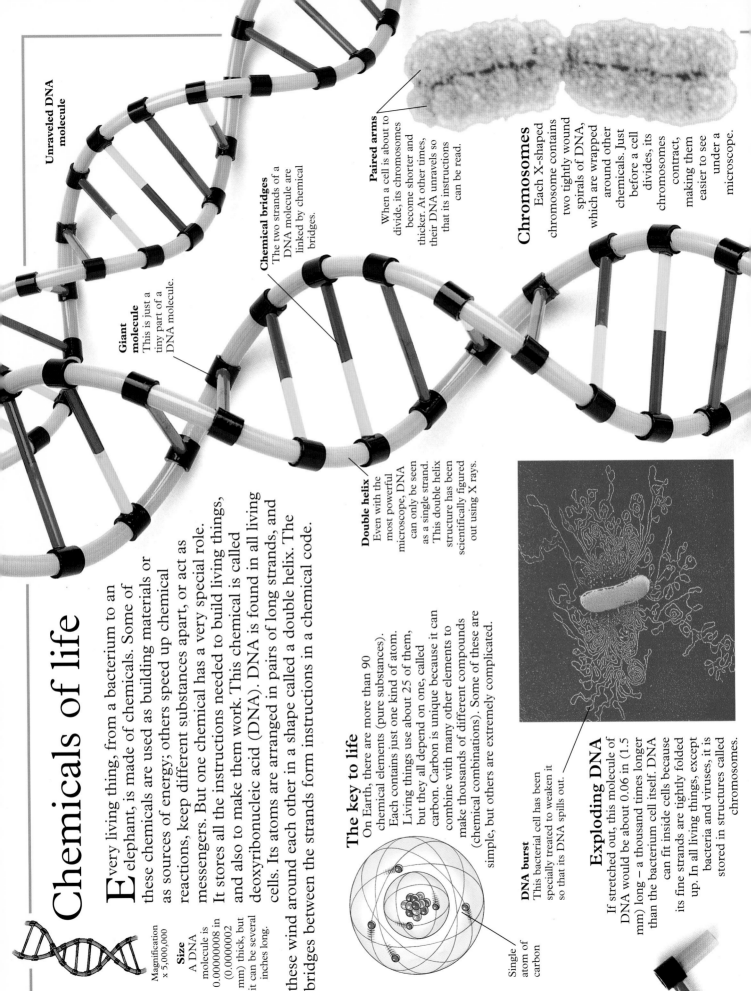

Unraveled DNA molecule

Chemicals of life

Every living thing, from a bacterium to an elephant, is made of chemicals. Some of these chemicals are used as building materials or as sources of energy; others speed up chemical reactions, keep different substances apart, or act as messengers. But one chemical has a very special role. It stores all the instructions needed to build living things, and also to make them work. This chemical is called deoxyribonucleic acid (DNA). DNA is found in all living cells. Its atoms are arranged in pairs of long strands, and these wind around each other in a shape called a double helix. The bridges between the strands form instructions in a chemical code.

Size
A DNA molecule is 0.00000008 in (0.000002 mm) thick, but it can be several inches long.

Magnification x 5,000,000

Giant molecule
This is just a tiny part of a DNA molecule.

Chemical bridges
The two strands of a DNA molecule are linked by chemical bridges.

Paired arms
When a cell is about to divide, its chromosomes become shorter and thicker. At other times, their DNA unravels so that its instructions can be read.

Chromosomes
Each X-shaped chromosome contains two tightly wound spirals of DNA, which are wrapped around other chemicals. Just before a cell divides, its chromosomes contract, making them easier to see under a microscope.

Double helix
Even with the most powerful microscope, DNA can only be seen as a single strand. This double helix structure has been scientifically figured out using X rays.

The key to life
On Earth, there are more than 90 chemical elements (pure substances). Each contains just one kind of atom. Living things use about 25 of them, but they all depend on one, called carbon. Carbon is unique because it can combine with many other elements to make thousands of different compounds (chemical combinations). Some of these are simple, but others are extremely complicated.

Single atom of carbon

DNA burst
This bacterial cell has been specially treated to weaken it so that its DNA spills out.

Exploding DNA
If stretched out, this molecule of DNA would be about 0.06 in (1.5 mm) long – a thousand times longer than the bacterium cell itself. DNA can fit inside cells because its fine strands are tightly folded up. In all living things, except bacteria and viruses, it is stored in structures called chromosomes.

The genetic code

The two strands of a DNA molecule are linked by chemical bridges. These are made up of four different substances, shown here in different colors. The bridges are arranged in long sequences. Each separate instruction in this sequence is a gene. The cell decodes these genetic instructions and uses them to make all the chemicals it needs.

Mine of information
A single molecule of DNA can contain more than a billion bridges – enough to spell out a huge number of instructions.

Genes
Each gene is made up of hundreds or thousands of bridges.

Human chromosomes
Human cells contain forty-six chromosomes.

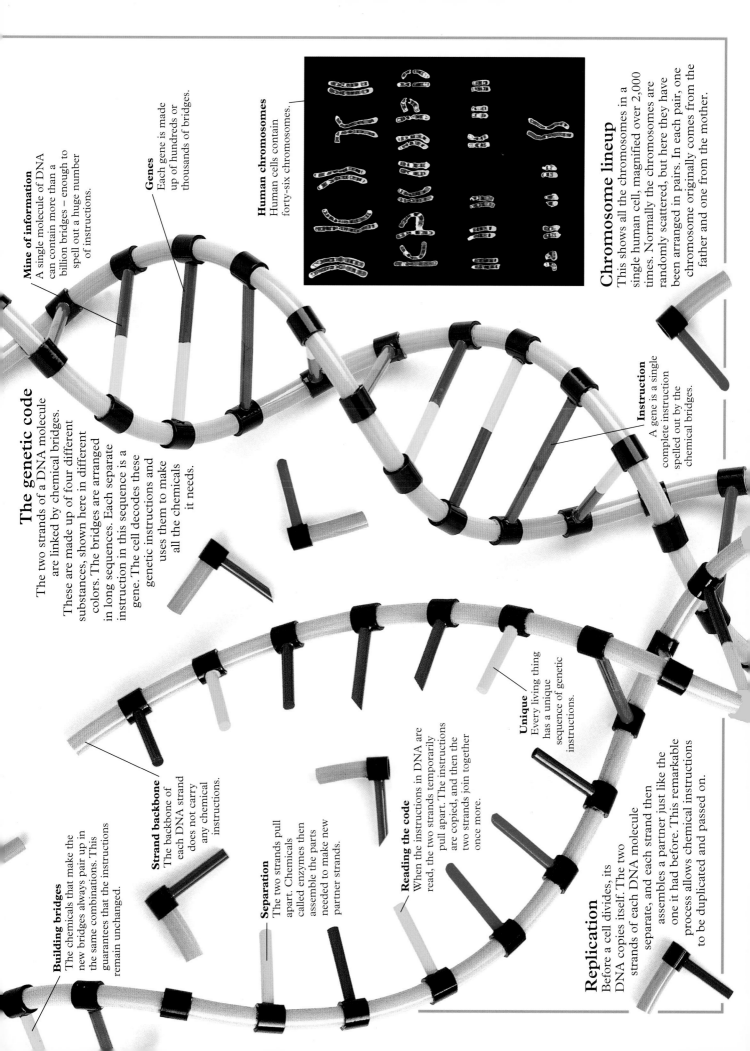

Chromosome lineup

This shows all the chromosomes in a single human cell, magnified over 2,000 times. Normally the chromosomes are randomly scattered, but here they have been arranged in pairs. In each pair, one chromosome originally comes from the father and one from the mother.

Instruction
A gene is a single complete instruction spelled out by the chemical bridges.

Building bridges
The chemicals that make the new bridges always pair up in the same combinations. This guarantees that the instructions remain unchanged.

Strand backbone
The backbone of each DNA strand does not carry any chemical instructions.

Separation
The two strands pull apart. Chemicals called enzymes then assemble the parts needed to make new partner strands.

Reading the code
When the instructions in DNA are read, the two strands temporarily pull apart. The instructions are copied, and then the two strands join together once more.

Unique
Every living thing has a unique sequence of genetic instructions.

Replication

Before a cell divides, its DNA copies itself. The two strands of each DNA molecule separate, and each strand then assembles a partner just like the one it had before. This remarkable process allows chemical instructions to be duplicated and passed on.

Life's building blocks

Size
Most animal cells are between 0.0002 and 0.0016 in (0.005 and 0.04 mm) across.

Cells are the tiny units that make up all living things. Some forms of life, such as bacteria, contain just a single cell. Many others consist of thousands or billions of cells living and working together. Although cells are incredibly varied, they have some common features. They are surrounded by a flexible membrane and contain a jellylike fluid called cytoplasm. In most living things except bacteria, the cytoplasm contains a variety of structures called organelles (little organs), which carry out many vital chemical processes.

Mitochondrion
To live, cells need energy. Most obtain it by breaking down chemicals using organelles called mitochondria. Each mitochondrion has a folded inner membrane, where energy is released, and a smooth outer membrane.

Outer membrane
This separates the mitochondrion from the cell.

Cell walls
Plant cells are surrounded by a rigid outer layer called a cell wall. The wall is made of fibers of a substance called cellulose, arranged in crisscrossing layers for extra strength. This picture shows some of these fibers magnified over 750 times.

Golgi body
This gathers substances made by the cell and carries them to the cell's surface.

Lysosomes
These bubblelike chambers contain digestive chemicals, which break down food substances or digest the cell itself.

Centrioles
Cells reproduce by dividing in two. The two centrioles help organize this process.

Mitochondria
These are the cell's powerhouses, where energy-releasing reactions occur.

Vesicles
These hollows absorb substances and travel into the cell.

Plasma membrane

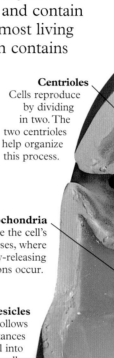

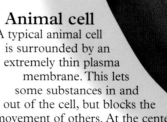

Mesophyll cells
These cells use the sun's energy to make food.

Working together
This magnified leaf slice shows several types of cells packed closely together. Each type performs a specific function. Some form a tough surface; others contain organelles called chloroplasts that harness the energy in sunlight; others ferry substances between the leaf and the rest of the plant.

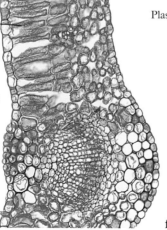

Animal cell
A typical animal cell is surrounded by an extremely thin plasma membrane. This lets some substances in and out of the cell, but blocks the movement of others. At the center of the cell is the nucleus – the all-important command center that contains the cell's DNA (pp.10–11). The cell is bathed in fluid that brings the cell nutrients and oxygen.

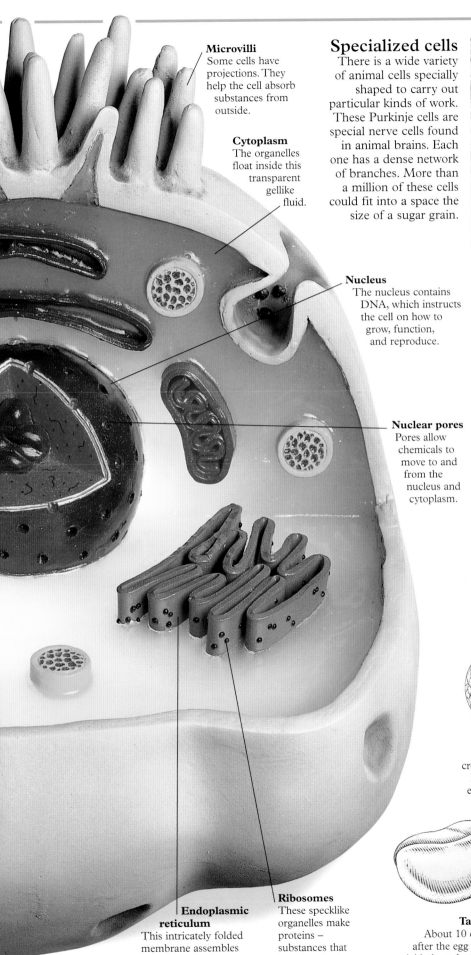

Microvilli
Some cells have projections. They help the cell absorb substances from outside.

Cytoplasm
The organelles float inside this transparent gellike fluid.

Specialized cells

There is a wide variety of animal cells specially shaped to carry out particular kinds of work. These Purkinje cells are special nerve cells found in animal brains. Each one has a dense network of branches. More than a million of these cells could fit into a space the size of a sugar grain.

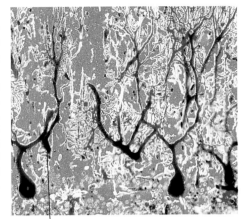

Branching network
Neighboring cells are connected by a dense network of branches.

Nucleus
The nucleus contains DNA, which instructs the cell on how to grow, function, and reproduce.

Nuclear pores
Pores allow chemicals to move to and from the nucleus and cytoplasm.

Growth and development

Most living things start life as a single cell. From this one cell, a complete organism gradually develops. Like most animals, a frog starts life as a fertilized egg, which is a giant single cell. Its development begins when the original cell divides in two. More divisions quickly follow, and the new cells develop to form the body's different parts.

Fertilized egg
The egg contains lots of yolk, which provides the food it needs while it develops.

Four-cell stage
After dividing into two, it divides into four cells. The cells look identical.

Blastula
After many divisions, thousands of cells form a hollow, fluid-filled ball called a blastula.

Gastrula
The ball folds in on itself, creating a gastrula with three layers of cells. The gastrula elongates and a groove runs along its upper surface.

Developing body parts
The layers of cells develop into different parts of the body, such as the head and internal organs.

Endoplasmic reticulum
This intricately folded membrane assembles proteins and other useful substances.

Ribosomes
These specklike organelles make proteins – substances that have hundreds of different uses.

Tadpole
About 10 days after the egg was laid, the tadpole is ready to swim away.

Virus attack

Size
A typical virus
is 0.00009 in
(0.000225 mm)
high.

Viruses are nature's strangest invaders. They cannot feed or grow, and most of the time they show no signs of life. But if a virus touches a suitable cell, all this suddenly changes. The virus locks onto the cell wall and injects a set of chemical instructions. These spread through the cell and seize command, forcing the cell to stop its normal work and make copies of the virus instead. As soon as the copies are complete, they spill out of the cell, often leaving it dead. There are thousands of kinds of viruses. Some are harmless; others cause deadly diseases. But all exist on the very edge of life.

Injecting DNA
Tail sheath contracts
and DNA is injected
into the bacterial cell.

Locking on
When a bacteriophage
touches a bacterial
cell, its tail fibers lock
onto the cell's surface.

Head
This contains
a strand of DNA.

Bacteriophage
This bizarre-looking virus attacks bacteria. It is so small that 5,000 laid end to end would only just reach across a pin point. Like all viruses, it contains a strand of nucleic acid – in this case DNA (deoxyribonucleic acid). The DNA contains all the instructions the virus needs to make copies of itself.

1 Once a bacteriophage is attached to a bacterium, it produces an enzyme (special protein) that dissolves an opening in the bacterium's cell wall. The virus then injects its DNA through this opening.

Short
pins on
baseplate

DNA strand
DNA enables
virus to make
copies of itself.

Tail fiber
The six tail fibers are also made of proteins. Their tips can automatically lock onto a bacterium's surface.

Tail fiber knee
Each tail fiber has a knee that bends on landing.

Tail sheath
This hollow tube can contract to inject the virus's DNA into a bacterium.

Influenza virus
Spikes lock onto the surface of the target cell.

Drifting bacteriophage
Viruses cannot move themselves. They drift through air and water, and meet bacteria by accident.

Empty cases
Their work done, the empty viral cases remain outside the bacterium.

Tobacco mosaic virus
Each mosaic virus consists of a strand of nucleic acid wound into a spiral.

Plant viruses
Viruses attack all kinds of living things, including plants. Rod-shaped mosaic viruses infect plants, such as tomatoes and tobacco. Mosaic viruses attack cells in leaves, sometimes causing the plant to die. Viruses are often spread from plant to plant by sap-sucking insects such as aphids.

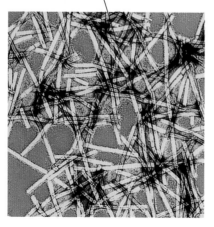

Rounded virus
This common virus causes influenza when it attacks the cells that line the throat. There are many varieties of flu viruses, each with a slightly different chemical makeup. This makes it difficult for our bodies to recognize these viruses and fight them off.

2 Once it is inside, the virus's DNA reprograms the bacterium. It stops it from carrying out its normal work, and forces it to make the parts needed for new viruses. These parts include new DNA. Each part of the new virus is made separately.

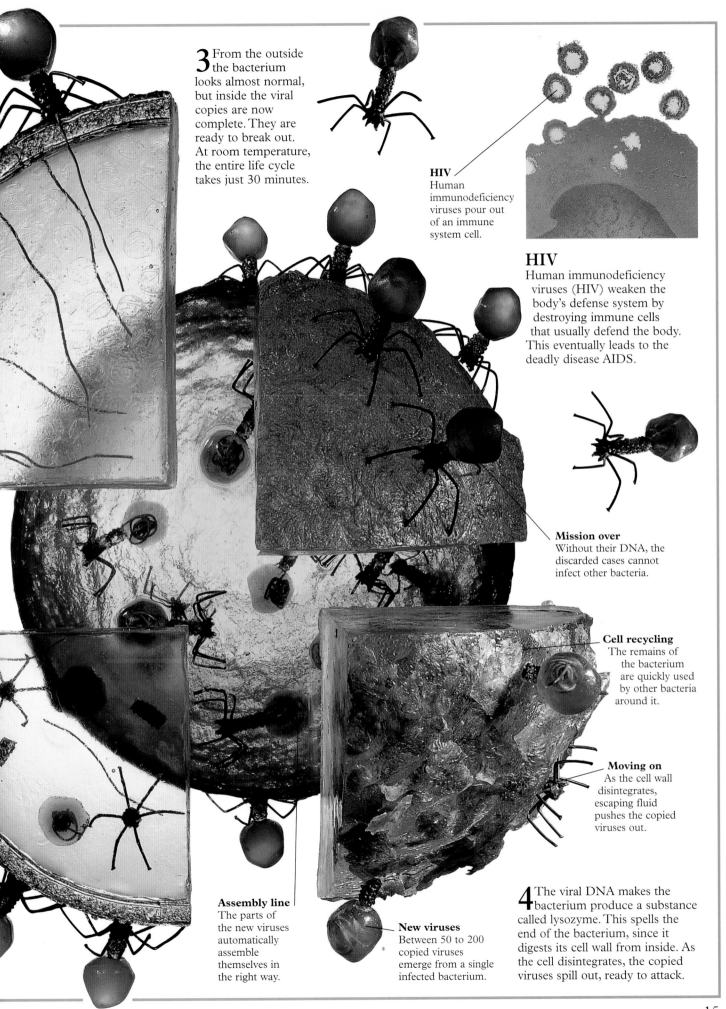

3 From the outside the bacterium looks almost normal, but inside the viral copies are now complete. They are ready to break out. At room temperature, the entire life cycle takes just 30 minutes.

HIV
Human immunodeficiency viruses pour out of an immune system cell.

HIV
Human immunodeficiency viruses (HIV) weaken the body's defense system by destroying immune cells that usually defend the body. This eventually leads to the deadly disease AIDS.

Mission over
Without their DNA, the discarded cases cannot infect other bacteria.

Cell recycling
The remains of the bacterium are quickly used by other bacteria around it.

Moving on
As the cell wall disintegrates, escaping fluid pushes the copied viruses out.

Assembly line
The parts of the new viruses automatically assemble themselves in the right way.

New viruses
Between 50 to 200 copied viruses emerge from a single infected bacterium.

4 The viral DNA makes the bacterium produce a substance called lysozyme. This spells the end of the bacterium, since it digests its cell wall from inside. As the cell disintegrates, the copied viruses spill out, ready to attack.

Teeming hordes

Bacteria are the smallest and most abundant forms of life on Earth. They are so tiny that several billion could fit in a matchbox, and they live everywhere, from the surface of our skin to tiny spaces deep within rocks. Many bacteria are vital for the survival of living things, but others cause diseases. Most bacteria survive by absorbing food substances from their surroundings, but some collect energy from sunshine. Bacteria are single-celled and usually reproduce by dividing in two. In favorable conditions, they can do this every 20 minutes.

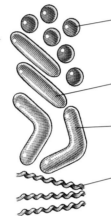

Magnification
x 25,000

Size
The average bacterium is about 0.00004 in (0.001 mm) long.

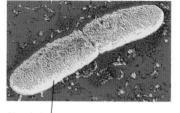

Rhizobium
This bacterium is fastening itself to a pea plant root.

The good...
The bacterium *Rhizobium* collects nitrogen (a vital gas) from the soil. It turns it into a form that growing plants can use.

...and the bad
The bacterium *Yersinia* causes the plague, a deadly disease. It often lives on rats and fleas, but can be passed on to people.

Yersinia
The plague killed millions of people in the Middle Ages.

Types of bacteria
Scientists have identified about 10,000 species of bacteria, and many more probably await discovery. Bacteria are often classified by their shape. However, since there is a limited range of shapes, scientists test bacteria chemically to determine the exact species. Each species has a different chemical makeup.

Coccus
Rounded cocci bacteria usually live separately, but some live in twos or fours.

Bacillus
Many bacteria are rod shaped.

Vibrio
A vibrio is a comma-shaped spiral bacterium.

Spirillum
This kind of spiral bacterium is shaped like a corkscrew.

Cross-wall
Wall is developing between the new cells.

Clump of DNA

Dividing
Before a bacterium can divide in two, it has to copy its DNA (pp.10–11). Here a *Staphylococcus* bacterium has done just that, and its DNA has formed two separate clumps. A cross-wall has grown across the cell, cutting it in two. When the cross-wall is complete, the two cells will split apart.

Living on light
Cyanobacteria (sometimes called blue-green algae) live by harnessing the energy in sunlight. They use it to build up their cells and to work. Plants also live like this, but cyanobacteria perfected this way of life long before plants first appeared. This cyanobacterium, called *Trichodesmium,* drifts in water.

Living together
Instead of living on their own, *Trichodesmium* cells live in filaments (slender strands) surrounded by slimy jelly. They have microscopic floats that keep them near water's sunlit surface.

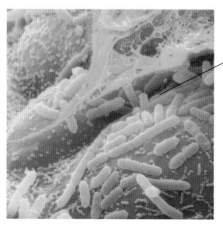

Bacteria on lining of nostril

Bacteria on the body
In the average human body, bacteria outnumber the body's own cells. Most live on the body's surfaces and help keep harmful bacteria at bay. However, if they manage to get inside the body's tissues, the same bacteria can trigger diseases. The bacterium *Haemophilus* lives on the lining of the mouth and nose.

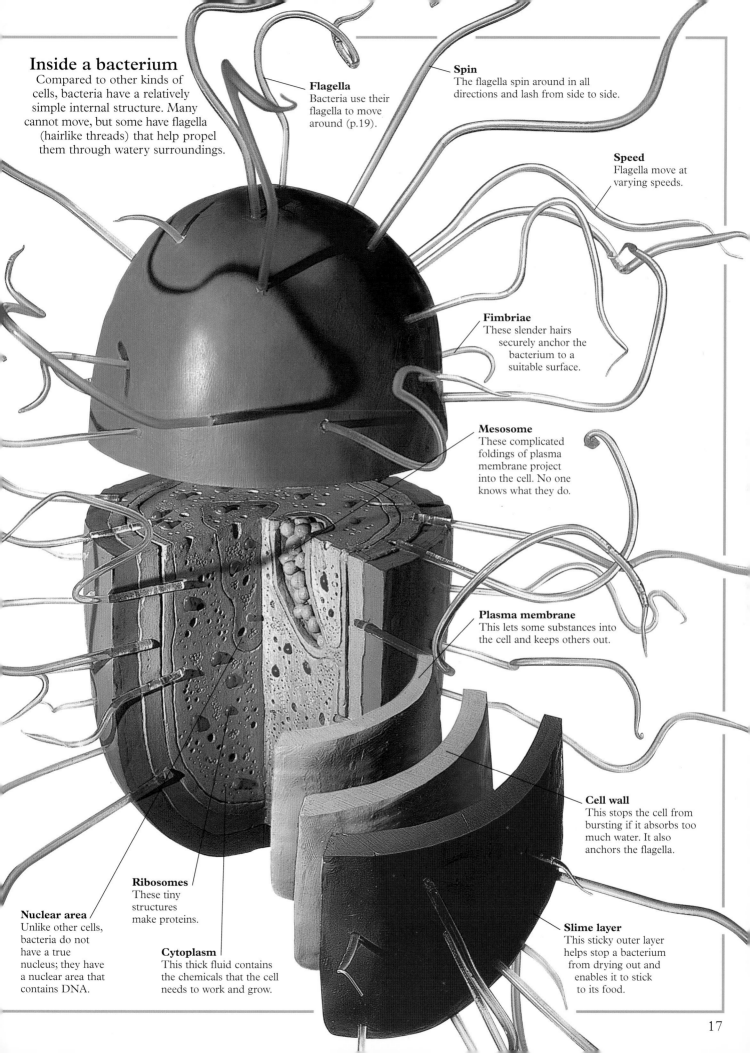

Inside a bacterium

Compared to other kinds of cells, bacteria have a relatively simple internal structure. Many cannot move, but some have flagella (hairlike threads) that help propel them through watery surroundings.

Flagella
Bacteria use their flagella to move around (p.19).

Spin
The flagella spin around in all directions and lash from side to side.

Speed
Flagella move at varying speeds.

Fimbriae
These slender hairs securely anchor the bacterium to a suitable surface.

Mesosome
These complicated foldings of plasma membrane project into the cell. No one knows what they do.

Plasma membrane
This lets some substances into the cell and keeps others out.

Cell wall
This stops the cell from bursting if it absorbs too much water. It also anchors the flagella.

Nuclear area
Unlike other cells, bacteria do not have a true nucleus; they have a nuclear area that contains DNA.

Ribosomes
These tiny structures make proteins.

Cytoplasm
This thick fluid contains the chemicals that the cell needs to work and grow.

Slime layer
This sticky outer layer helps stop a bacterium from drying out and enables it to stick to its food.

Microbes on the move

Size
An *Euglena*'s cell is about 0.002 in (0.05 mm) long.

Not all microbes can move, but some scurry from place to place. Many move to find food or to approach light, so they can harness the energy that it contains. Some microscopic animals, such as tardigrades (pp.40–41), move by using their legs. However, the smallest microbes do not have any legs and move in other ways. Many of them beat tiny hairs that work like oars or propellers. A few crawl along by squeezing and stretching their cells. Some of these microbes move very slowly, but others can speed along at more than 6 in (15 cm) a minute.

Body first, flagellum second
When *Euglena* crawls forward, its flagellum trails behind its body.

Nucleus of cell

Short flagellum

Rowing along
Euglena is a single-celled microbe that lives in ponds, puddles, and seashore mud. It lives mainly like a plant, by soaking up the energy in sunlight. *Euglena* needs to stay near the surface, where the light is brightest. It swims through the water by lashing its long, hairlike flagellum.

Flagellum
This hairlike thread beats from side to side. Unlike most flagella, it pulls the cell along.

Flagellar pocket
Euglena's flagella are anchored inside a pocket.

Ciliate
Cilia are arranged in rows and move in waves.

Eyespot
Euglena's eyespot guides the cell toward the light.

Sliding fibers
Fibers inside the flagellum slide over each other to make it move.

Unwelcome guest
The pear-shaped *Giardia* lives by absorbing some of the food that its host has eaten.

Power pack
Two bundles of four flagella propel *Giardia* along.

Swimming with cilia
Instead of swimming by beating flagella, some microorganisms use many thousands of much smaller hairs called cilia. Cilia often cover the whole of the cell, and they beat in sequence. This ciliate cell, *Tetrahymena*, uses its cilia to steer and change direction as it speeds through ponds and streams. Since the ciliate is so small, it halts as soon as its cilia stop beating.

Swimming inside animals
The single-celled parasite *Giardia* lives in water and in the intestines of animals and people. It swims by waving eight flagella, and it fastens itself in place with a special sticky disk. People often become infected with *Giardia* by drinking contaminated water.

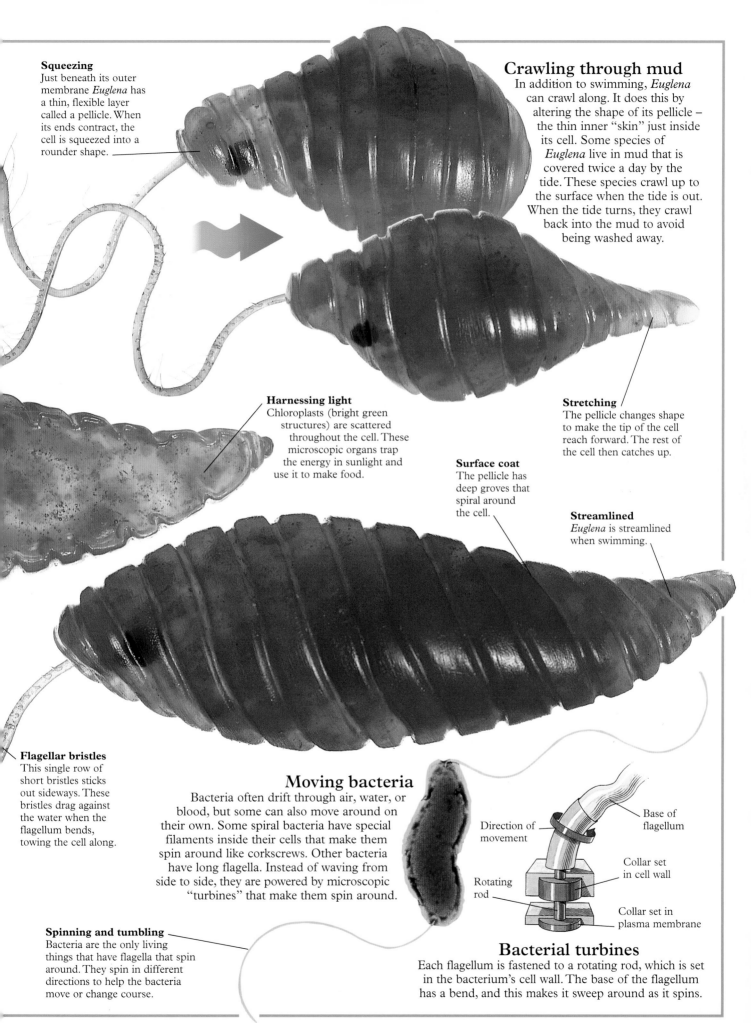

Squeezing
Just beneath its outer membrane *Euglena* has a thin, flexible layer called a pellicle. When its ends contract, the cell is squeezed into a rounder shape.

Crawling through mud
In addition to swimming, *Euglena* can crawl along. It does this by altering the shape of its pellicle – the thin inner "skin" just inside its cell. Some species of *Euglena* live in mud that is covered twice a day by the tide. These species crawl up to the surface when the tide is out. When the tide turns, they crawl back into the mud to avoid being washed away.

Harnessing light
Chloroplasts (bright green structures) are scattered throughout the cell. These microscopic organs trap the energy in sunlight and use it to make food.

Stretching
The pellicle changes shape to make the tip of the cell reach forward. The rest of the cell then catches up.

Surface coat
The pellicle has deep groves that spiral around the cell.

Streamlined
Euglena is streamlined when swimming.

Flagellar bristles
This single row of short bristles sticks out sideways. These bristles drag against the water when the flagellum bends, towing the cell along.

Moving bacteria
Bacteria often drift through air, water, or blood, but some can also move around on their own. Some spiral bacteria have special filaments inside their cells that make them spin around like corkscrews. Other bacteria have long flagella. Instead of waving from side to side, they are powered by microscopic "turbines" that make them spin around.

Direction of movement

Base of flagellum

Rotating rod

Collar set in cell wall

Collar set in plasma membrane

Spinning and tumbling
Bacteria are the only living things that have flagella that spin around. They spin in different directions to help the bacteria move or change course.

Bacterial turbines
Each flagellum is fastened to a rotating rod, which is set in the bacterium's cell wall. The base of the flagellum has a bend, and this makes it sweep around as it spins.

Shape changers

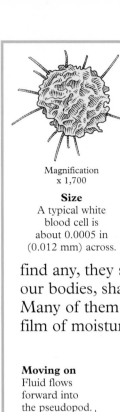

Magnification
x 1,700

Size
A typical white blood cell is about 0.0005 in (0.012 mm) across.

Although you cannot feel it, some of your cells are constantly changing shape. White blood cells wander through your body in search of microscopic invaders that could cause infection. If they find any, they swallow them up. In the world outside our bodies, shape-changing cells are quite common. Many of them creep through water, or through the thin film of moisture that surrounds individual particles of soil.

Secret signals
Bacteria release chemicals that are not normally found in the body. This helps white blood cells track them down.

Squeezing flat
By flattening itself, the white blood cell can squeeze through the lining of small blood vessels.

Moving on
Fluid flows forward into the pseudopod.

Speed ahead
At top speed, an amoeba can move at about 1 in (2.5 cm) an hour.

Pseudopod
This spreads forward in the direction of travel or toward food.

Inner fluid

Outer jelly

Vacuole
This bubble-like chamber stores food.

Rounding up the enemy

White blood cells ride around in your blood, but they also break out of your bloodstream and creep through the rest of your body. They can make themselves very thin to squeeze past other cells as they home in on bacteria or viruses. This white blood cell has located a group of bacteria. By changing shape, it is reaching out to engulf the bacteria and destroy them. During the fight against infection, millions of white blood cells often die, creating a white fluid called pus.

Pincer movement
Two or more pseudopods surround the amoeba's food.

Taken aboard
The amoeba engulfs its food and stores it in a vacuole (bubblelike chamber), where it is digested.

Reaching out
The white blood cell produces pseudopods (outgrowths), which reach out toward the bacteria.

Food in vacuole

Final moments
Once they have surrounded some bacteria, the pseudopods meet and join together. The bacteria are swallowed up into the cell and digested.

Tip of pseudopod

Flowing ahead

Single-celled amoebas live in ponds, damp soil, and other wet places. The cell's interior is made of a thick fluid, while the outer part is made of a firm jelly. In order to move, an amoeba converts some parts of the jelly into fluid. This flows forward in outgrowths called pseudopods, or false feet. When the rest of the cell catches up, new pseudopods form, and the cycle begins again.

Sticky grasp
Pseudopods are covered with gluelike chemicals. When they touch a bacterium, they usually stick fast.

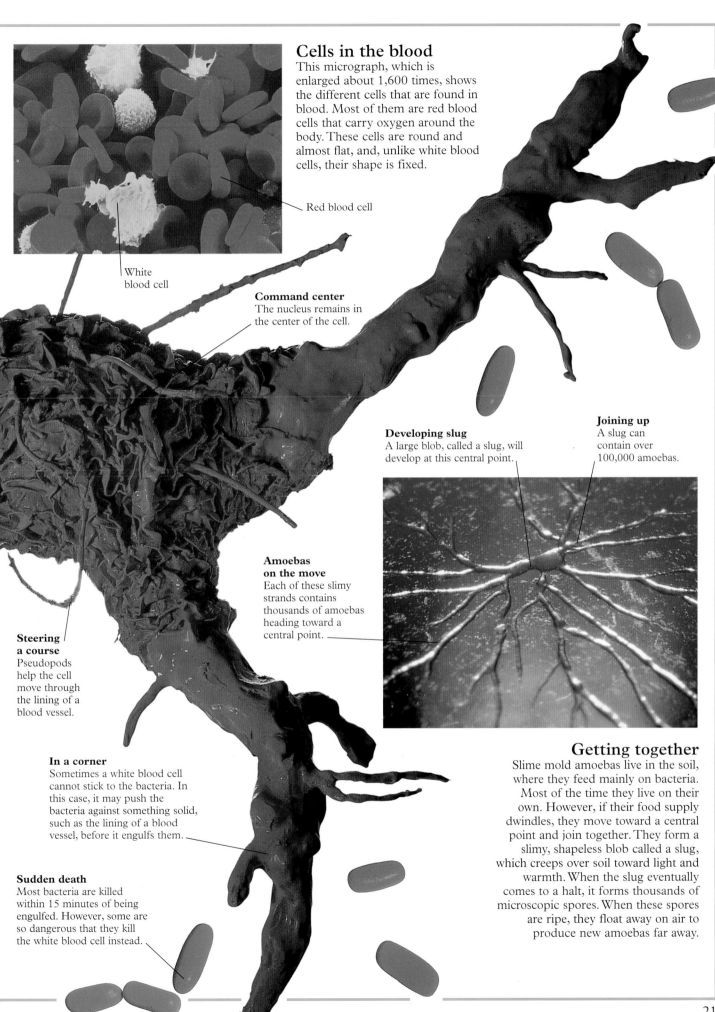

Cells in the blood

This micrograph, which is enlarged about 1,600 times, shows the different cells that are found in blood. Most of them are red blood cells that carry oxygen around the body. These cells are round and almost flat, and, unlike white blood cells, their shape is fixed.

— Red blood cell

White blood cell

Command center
The nucleus remains in the center of the cell.

Developing slug
A large blob, called a slug, will develop at this central point.

Joining up
A slug can contain over 100,000 amoebas.

Amoebas on the move
Each of these slimy strands contains thousands of amoebas heading toward a central point.

Steering a course
Pseudopods help the cell move through the lining of a blood vessel.

In a corner
Sometimes a white blood cell cannot stick to the bacteria. In this case, it may push the bacteria against something solid, such as the lining of a blood vessel, before it engulfs them.

Sudden death
Most bacteria are killed within 15 minutes of being engulfed. However, some are so dangerous that they kill the white blood cell instead.

Getting together

Slime mold amoebas live in the soil, where they feed mainly on bacteria. Most of the time they live on their own. However, if their food supply dwindles, they move toward a central point and join together. They form a slimy, shapeless blob called a slug, which creeps over soil toward light and warmth. When the slug eventually comes to a halt, it forms thousands of microscopic spores. When these spores are ripe, they float away on air to produce new amoebas far away.

The drifters

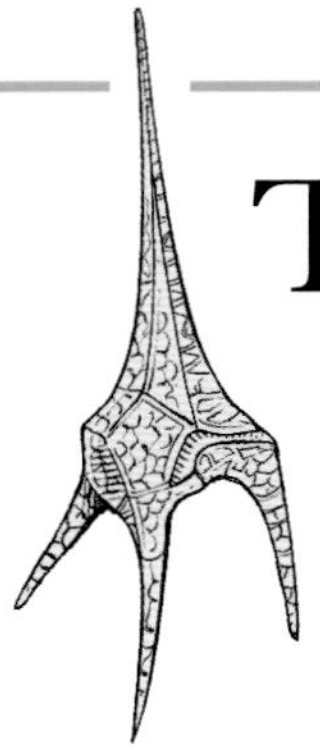

With its jagged spines and armor plating, *Ceratium* looks like something from another world. But it is actually a single-celled alga – one of thousands of kinds that drift in freshwater and the sea. Algae are among the most important living things on Earth. Like plants, they grow by harnessing the energy in sunlight. Although most are extremely small, they exist in vast numbers and form the basis of almost all marine food webs. Together, algae make up a mass of floating life called phytoplankton. In some parts of the sea, phytoplankton is quite sparse, but in others it is so dense that satellites can detect it from space.

Magnification
x 100

Size
A *Ceratium* cell measures up to 0.02 in (0.5 mm) from top to bottom.

Living together

This common colonial alga, *Volvox*, lives in ponds and ditches. Each colony is up to 0.04 in (1 mm) across and contains several hundred cells, set in a sphere of jelly. Each cell has two flagella (whiplike hairs), and when they beat, the entire colony tumbles through the water. Many of the colonies shown here contain offspring colonies.

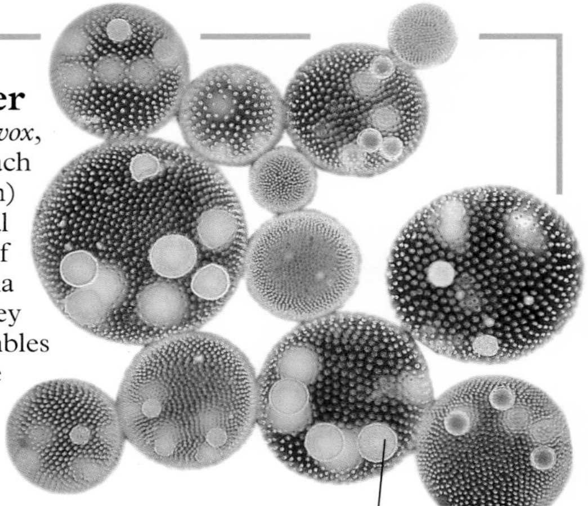

Offspring colonies
The offspring colonies tumble inside the parent colony.

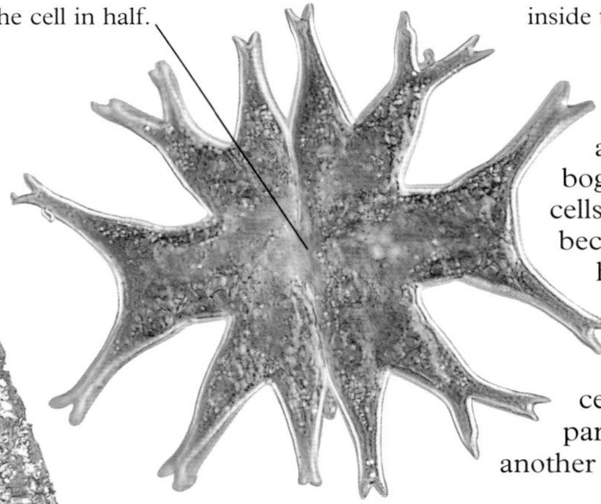

Desmids

This bright green alga lives in ponds and boggy ground. Desmid cells are easy to recognize because they look like they have been cut in two. When a desmid reproduces, the two halves of the cell separate. Each part then grows another opposite half.

Dividing line
A clear line cuts the cell in half.

Spinning around
The shallow groove that runs around the middle of the cell contains one of the two flagella. When this flagellum moves, the whole cell spins around.

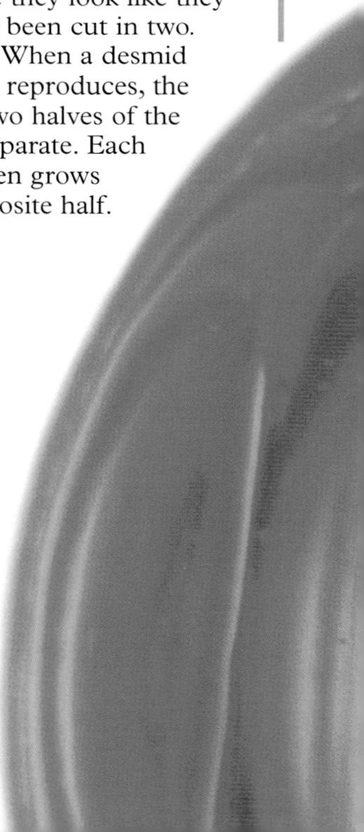

The world of algae

Ceratium belongs to a group of algae called dinoflagellates, which are most common in tropical seas. When conditions are ideal for growth, they sometimes form a dense layer near the water's surface, staining it brown or red. *Ceratium* is harmless, but some dinoflagellates produce powerful poisons. This "red tide" contains billions of these dangerous algae; their poison often kills fish.

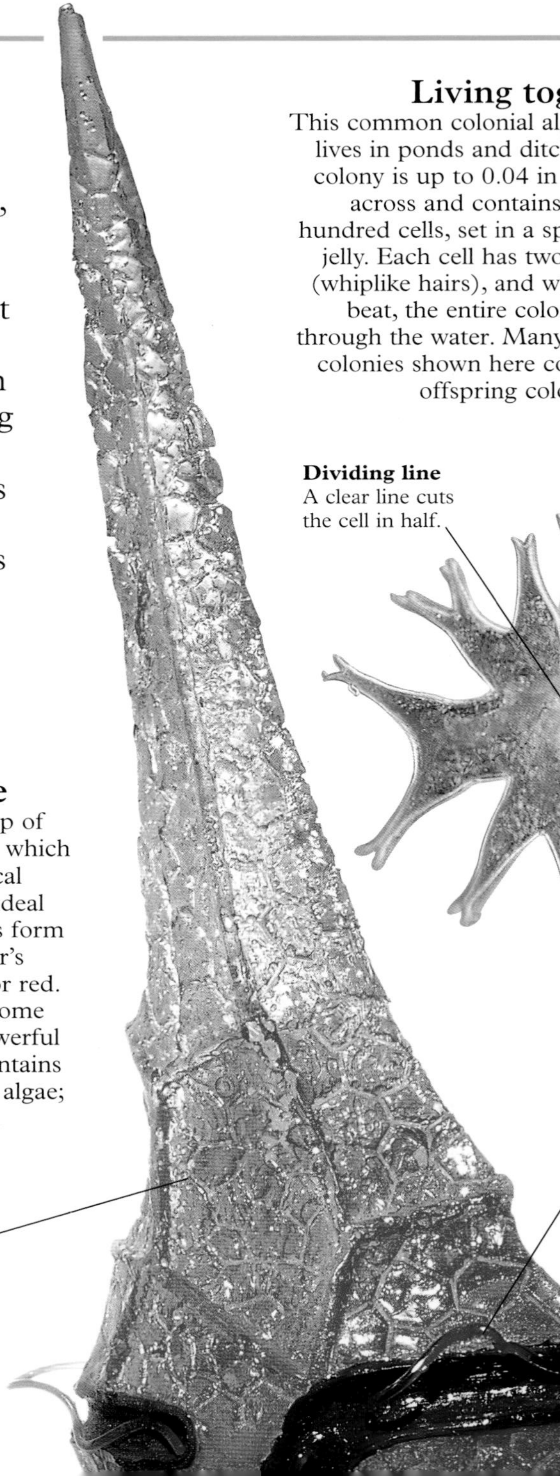

Algal overgrowth
Many algae undergo annual population explosions or blooms.

Chlorophyll
Ceratium contains chlorophyll, which is a green pigment that traps energy from sunlight. It also contains other pigments that mask the chlorophyll, giving it a reddish-brown color.

Air bubble

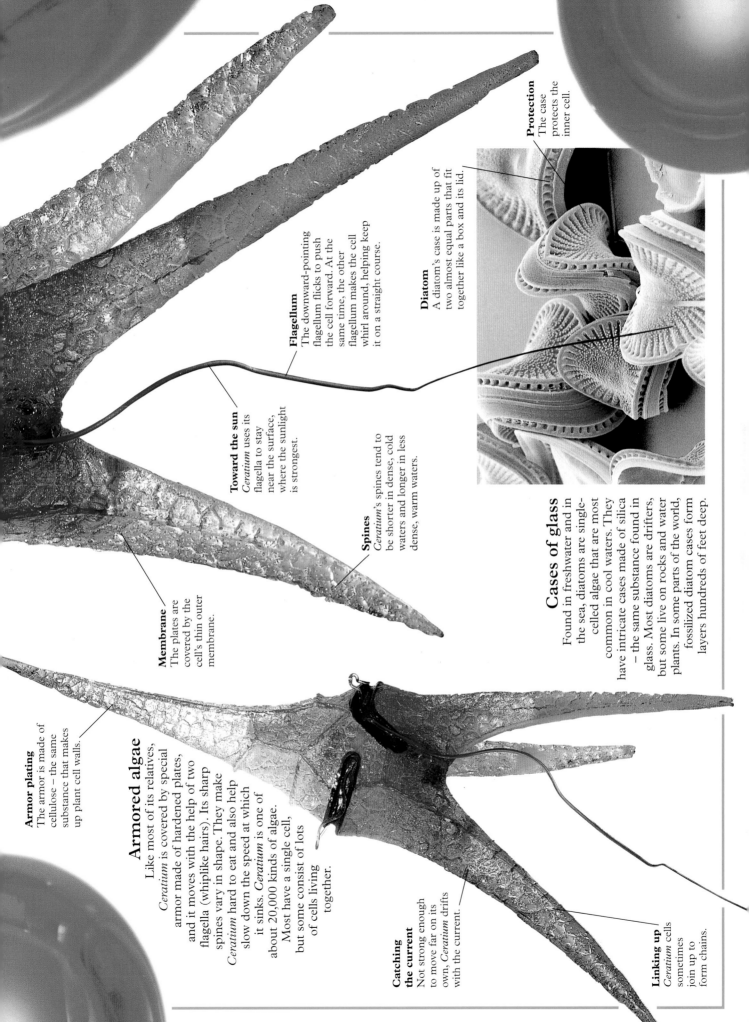

Protection
The case protects the inner cell.

Diatom
A diatom's case is made up of two almost equal parts that fit together like a box and its lid.

Flagellum
The downward-pointing flagellum flicks to push the cell forward. At the same time, the other flagellum makes the cell whirl around, helping keep it on a straight course.

Toward the sun
Ceratium uses its flagella to stay near the surface, where the sunlight is strongest.

Spines
Ceratium's spines tend to be shorter in dense, cold waters and longer in less dense, warm waters.

Membrane
The plates are covered by the cell's thin outer membrane.

Cases of glass
Found in freshwater and in the sea, diatoms are single-celled algae that are most common in cool waters. They have intricate cases made of silica – the same substance found in glass. Most diatoms are drifters, but some live on rocks and water plants. In some parts of the world, fossilized diatom cases form layers hundreds of feet deep.

Armor plating
The armor is made of cellulose – the same substance that makes up plant cell walls.

Armored algae
Like most of its relatives, *Ceratium* is covered by special armor made of hardened plates, and it moves with the help of two flagella (whiplike hairs). Its sharp spines vary in shape. They make *Ceratium* hard to eat and also help slow down the speed at which it sinks. *Ceratium* is one of about 20,000 kinds of algae. Most have a single cell, but some consist of lots of cells living together.

Catching the current
Not strong enough to move far on its own, *Ceratium* drifts with the current.

Linking up
Ceratium cells sometimes join up to form chains.

Water-drop world

Magnification
x 150

Size
The average
Vorticella is 0.008
in (0.2 mm) long.

Over three centuries ago, when scientists first examined water under microscopes, they were amazed by what they saw. They discovered that drops of pond water or seawater teem with miniature forms of life, many of them on the move. They called these microorganisms "animalcules" (tiny animals). Today they are known as protozoans. Although they are only single-celled, they show many different types of behavior and lead quite complicated lives. They often behave like animals, and live by eating food. Some pursue prey like tiny hunters, while many just float near the surface. Others, like *Vorticella*, stay fastened in one place and wait for food to come within reach.

Cilia
Cilia (tiny hairs)
on the *Vorticella*'s
cup trap bacteria
and food particles.

Funnel
An opening on
the side of the
funnel (cup)
leads to
a mouth.

Beating cilia
Cilia beat as
it feeds.

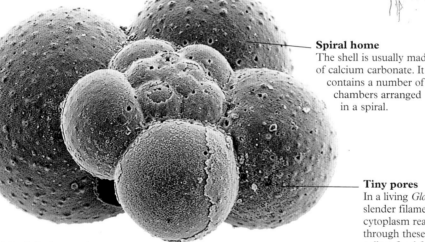

Spiral home
The shell is usually made
of calcium carbonate. It
contains a number of
chambers arranged
in a spiral.

Tiny pores
In a living *Globergerina*,
slender filaments of
cytoplasm reach out
through these pores to
collect food from the
surrounding water.

Multichambered shell
This beautiful shell is made by a protozoan called *Globigerina*,
which lives in the sea. *Globigerina* belongs to a group of protozoans
called foraminiferans. Like all its relatives, its shell is riddled with
pores. *Globigerina* is extremely abundant. In some places, its dead
shells pile up on the seabed to form a deep layer of ooze (fine mud).

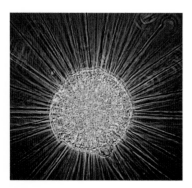

Living suns
Heliozoans, or "sun animals,"
have a round body with
radiating spikes that look like
rays of the sun. Each spike is
covered with sticky cytoplasm
(p.12), which carries food
toward the center of the cell.
Heliozoans live in freshwater
and in the sea. Some species
have slender stalks and are
anchored to the seabed.

Stalk
The stalk
contains a
special strand
that can contract,
making the stalk coil up.

Stable surface
Clusters of *Vorticella* fasten
themselves to stones, submerged
twigs, and animal shells.

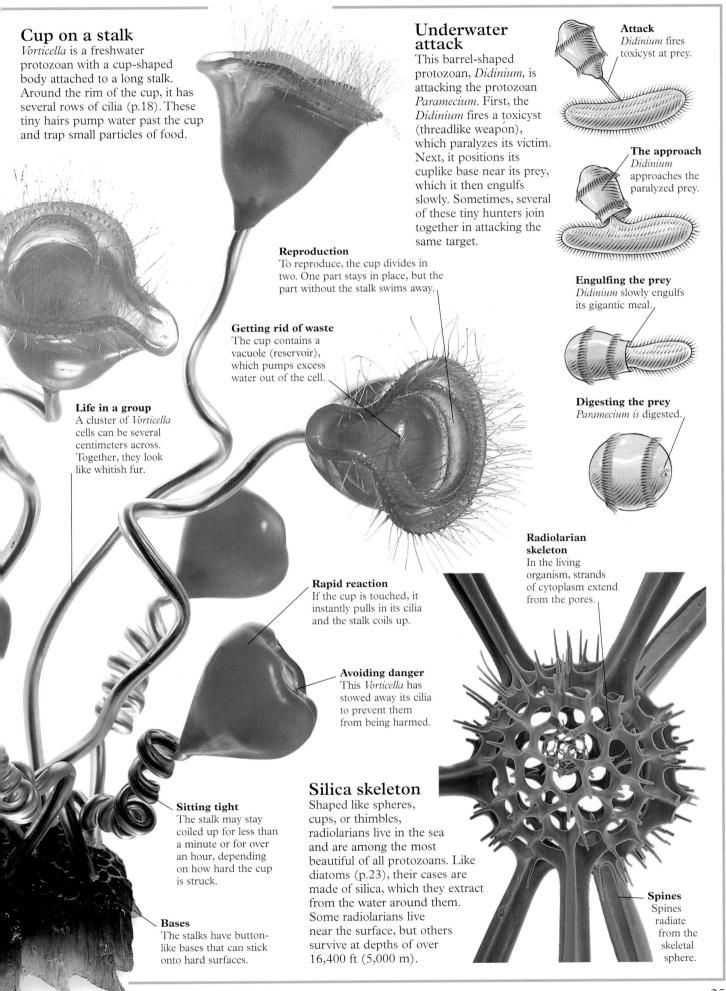

Cup on a stalk

Vorticella is a freshwater protozoan with a cup-shaped body attached to a long stalk. Around the rim of the cup, it has several rows of cilia (p.18). These tiny hairs pump water past the cup and trap small particles of food.

Life in a group
A cluster of *Vorticella* cells can be several centimeters across. Together, they look like whitish fur.

Reproduction
To reproduce, the cup divides in two. One part stays in place, but the part without the stalk swims away.

Getting rid of waste
The cup contains a vacuole (reservoir), which pumps excess water out of the cell.

Rapid reaction
If the cup is touched, it instantly pulls in its cilia and the stalk coils up.

Avoiding danger
This *Vorticella* has stowed away its cilia to prevent them from being harmed.

Sitting tight
The stalk may stay coiled up for less than a minute or for over an hour, depending on how hard the cup is struck.

Bases
The stalks have button-like bases that can stick onto hard surfaces.

Underwater attack

This barrel-shaped protozoan, *Didinium*, is attacking the protozoan *Paramecium*. First, the *Didinium* fires a toxicyst (threadlike weapon), which paralyzes its victim. Next, it positions its cuplike base near its prey, which it then engulfs slowly. Sometimes, several of these tiny hunters join together in attacking the same target.

Attack
Didinium fires toxicyst at prey.

The approach
Didinium approaches the paralyzed prey.

Engulfing the prey
Didinium slowly engulfs its gigantic meal.

Digesting the prey
Paramecium is digested.

Radiolarian skeleton
In the living organism, strands of cytoplasm extend from the pores.

Silica skeleton

Shaped like spheres, cups, or thimbles, radiolarians live in the sea and are among the most beautiful of all protozoans. Like diatoms (p.23), their cases are made of silica, which they extract from the water around them. Some radiolarians live near the surface, but others survive at depths of over 16,400 ft (5,000 m).

Spines
Spines radiate from the skeletal sphere.

Predators in plankton

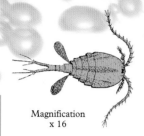

Size
A female *Cyclops* typically measures about 0.08 in (2 mm) long.

Magnification x 16

With a flick of her feathery antennae, a female *Cyclops* swims through pond water, carrying two egg sacs with her. Beneath her head she has specially shaped mouthparts, which she uses to seize tiny prey. Her antennae are very sensitive to vibrations, helping her home in on food. *Cyclops* and its relatives are only just big enough to be visible, but they are among the most numerous animals on Earth. They live in almost any kind of freshwater, from lakes and ponds to tree-hole pools, and billions of them thrive in the sea. These miniature water animals are crustaceans (relatives of crabs and lobsters). Along with other tiny drifting creatures, they usually live just beneath the water surface and form a mass of minute animals called zooplankton. They eat vast amounts of microscopic life. In turn, they provide food for many larger animals – and also for each other.

Zooplankton

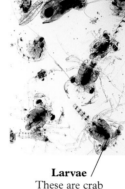

The animal part of plankton, zooplankton, contains the young of many crustaceans and fish, known as larvae. They change shape completely as they grow up. All crustaceans have exoskeletons (hard outer cases), which they shed periodically.

Larvae
These are crab and lobster larvae.

Tail bristles
Streamlined tail bristles give *Cyclops* extra speed when it swims.

Tail

Feathery antennae
These are used for swimming.

Antennae
All crustaceans have two pairs of antennae.

Body case
All of the body, except the head, is enclosed in a shieldlike case with a groove in the front.

Predator on patrol

Cyclops belongs to a group of crustaceans called copepods, which includes about 8,500 known species. Some of them filter tiny particles of food out of the water, but many also catch animal prey. *Cyclops* has a single eye, but its eyesight is very poor, so it finds its prey mostly by touch.

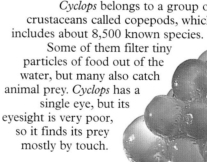

Abdomen
This is made up of five segments, which meet at flexible joints.

Egg sac
The female carries the eggs until they are ready to hatch. Each sac can hold up to 50 eggs.

Water fleas

Despite their name, water fleas are not insects, but tiny crustaceans. Like *Cyclops*, they swim by flicking their feathery antennae and move in rapid jerks. Male water fleas are much less common than female ones. In spring and summer, many females reproduce by growing eggs that do not need to be fertilized. The eggs hatch in a chamber inside the mother's body.

Rowing antennae

Young water flea

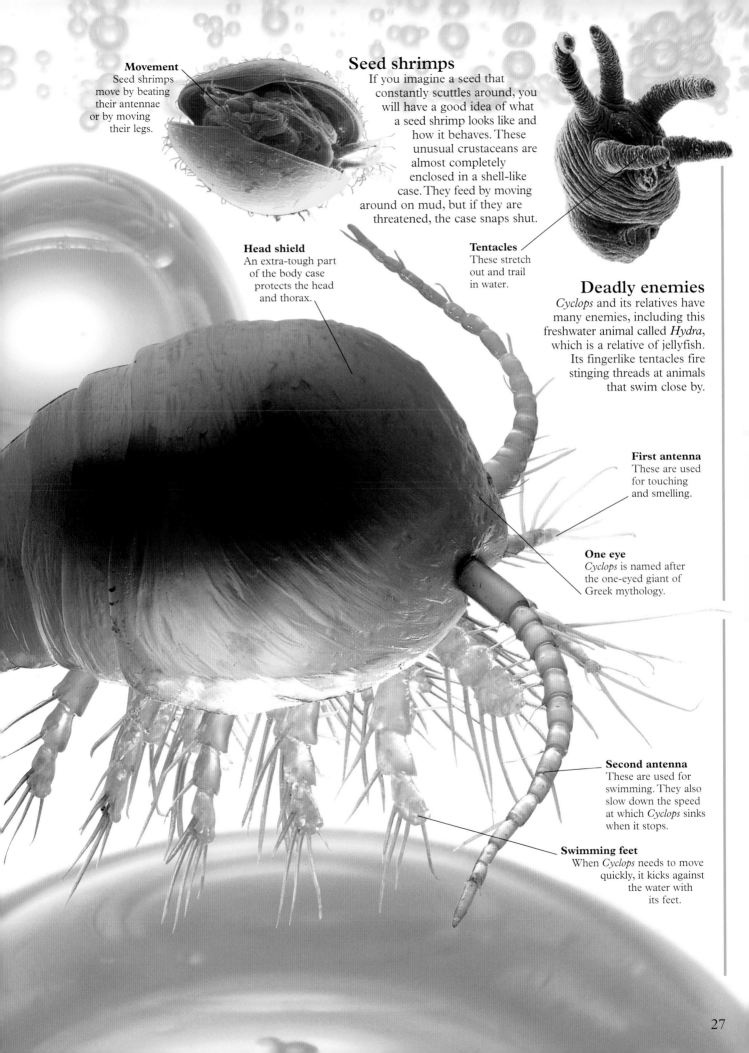

Movement
Seed shrimps move by beating their antennae or by moving their legs.

Seed shrimps

If you imagine a seed that constantly scuttles around, you will have a good idea of what a seed shrimp looks like and how it behaves. These unusual crustaceans are almost completely enclosed in a shell-like case. They feed by moving around on mud, but if they are threatened, the case snaps shut.

Tentacles
These stretch out and trail in water.

Head shield
An extra-tough part of the body case protects the head and thorax.

Deadly enemies

Cyclops and its relatives have many enemies, including this freshwater animal called *Hydra*, which is a relative of jellyfish. Its fingerlike tentacles fire stinging threads at animals that swim close by.

First antenna
These are used for touching and smelling.

One eye
Cyclops is named after the one-eyed giant of Greek mythology.

Second antenna
These are used for swimming. They also slow down the speed at which *Cyclops* sinks when it stops.

Swimming feet
When *Cyclops* needs to move quickly, it kicks against the water with its feet.

Floating on air

Magnification x 460

Size
Most airborne pollen grains are less than 0.002 in (0.05 mm) across.

If you climb a mountain and take a deep breath, you will breathe air and very little else. But on lower ground, the air is not so pure. It often contains smoke and dust, as well as millions of microscopic pollen grains and spores. Pollen grains are released by the male parts of flowers, while spores are released by fungi and bacteria. They drift far on the wind and enable plants to reproduce. However, when we breathe them in, they sometimes trigger an unpleasant reaction, such as the allergy hay fever.

Pine pollen

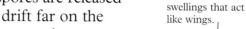

Hardened surface
The exine (outer wall) is very tough. If the grain gets buried in the ground, it can remain intact for thousands of years.

Wings
Pine pollen grains have two round swellings that act like wings.

Clouds of pollen
Fir tree releases clouds of yellow pollen.

Hazardous journey
This fir tree is shedding its powdery pollen into the air. At the mercy of the wind, only a tiny proportion of these minute grains will reach the right destination. To make up for this, wind-pollinated plants produce vast amounts of pollen.

Setting sail
The swellings expand as the grain matures.

Spiky surface
These spikes increase the pollen grain's surface area, which helps it get swept along by the wind.

Cluster of spores
The fruiting body contains masses of spores that are released into the air.

Intricately sculpted surface

Fungal spores
This fungus lives on plant remains. Like bread mold (pp.34–35), it forms its spores on the top of slender stalks, which gives the spores the best chance of being carried away by the wind. Spores have a simpler structure than pollen grains; they often consist of just a single cell. Each one is protected by a hard outer coat that enables it to survive harsh conditions.

Furrow
There are tiny pores inside this furrow.

Airborne pollen

Since plants cannot move, pollen brings their male and female sex cells together in order to create seeds. Each pollen grain contains male sex cells. If a pollen grain lands on the female part of a suitable flower, it fertilizes the flower. Many plants rely on animals to spread their pollen but some, such as grasses and conifers, use the wind. Each plant species has distinctively shaped and patterned pollen grains.

Filament of stigma

Ragweed pollen grains

A pollen grain is very light so it can float.

Fertilization

Above, ragweed pollen grains have successfully landed on the stigma (female part) of a ragweed flower. The stigma has many filaments on which the pollen can land. The interior of each pollen grain contains two or three male nuclei. These travel down a tube to fertilize the flower's female cells.

Antennae
A thrips uses its antennae to home in on its food plants.

Body size
A thrips's body is often just 0.08 in (2 mm) long.

Thrips

Ragweed pollen

Feathery wings
Thrips have narrow wings with long fringes.

Family feature
Most species of the daisy family, which includes ragweed, have spiky pollen.

Willow pollen

Drifters

In spring and summer, huge clouds of tiny insects, such as thrips and aphids, travel through the air. They all have wings, but they are so small they make little headway on their own. Instead, they ride on currents of air. They get on board by flying upward from plants, and they get off again by closing their wings and drifting back to the ground.

Germ pore
If a pollen grain lands on a flower, a slender tube will grow out of one of its pores to reach a female cell.

29

Dust dwellers

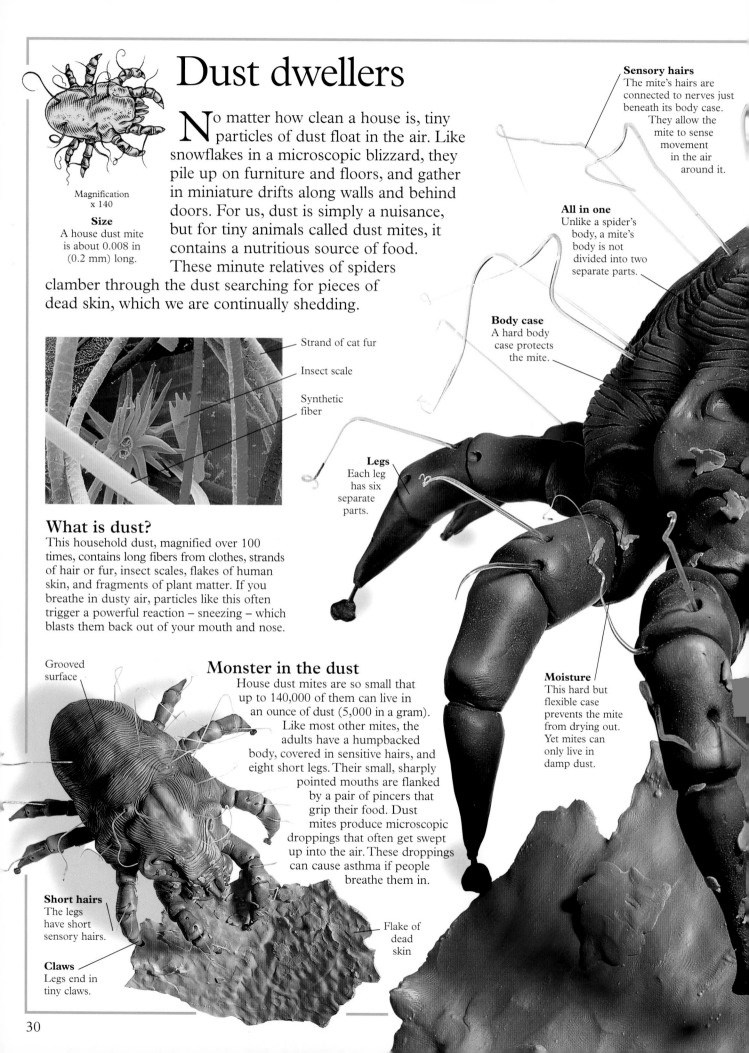

No matter how clean a house is, tiny particles of dust float in the air. Like snowflakes in a microscopic blizzard, they pile up on furniture and floors, and gather in miniature drifts along walls and behind doors. For us, dust is simply a nuisance, but for tiny animals called dust mites, it contains a nutritious source of food. These minute relatives of spiders clamber through the dust searching for pieces of dead skin, which we are continually shedding.

Magnification
x 140

Size
A house dust mite is about 0.008 in (0.2 mm) long.

Strand of cat fur

Insect scale

Synthetic fiber

What is dust?
This household dust, magnified over 100 times, contains long fibers from clothes, strands of hair or fur, insect scales, flakes of human skin, and fragments of plant matter. If you breathe in dusty air, particles like this often trigger a powerful reaction – sneezing – which blasts them back out of your mouth and nose.

Grooved surface

Monster in the dust
House dust mites are so small that up to 140,000 of them can live in an ounce of dust (5,000 in a gram). Like most other mites, the adults have a humpbacked body, covered in sensitive hairs, and eight short legs. Their small, sharply pointed mouths are flanked by a pair of pincers that grip their food. Dust mites produce microscopic droppings that often get swept up into the air. These droppings can cause asthma if people breathe them in.

Short hairs
The legs have short sensory hairs.

Claws
Legs end in tiny claws.

Sensory hairs
The mite's hairs are connected to nerves just beneath its body case. They allow the mite to sense movement in the air around it.

All in one
Unlike a spider's body, a mite's body is not divided into two separate parts.

Body case
A hard body case protects the mite.

Legs
Each leg has six separate parts.

Moisture
This hard but flexible case prevents the mite from drying out. Yet mites can only live in damp dust.

Flake of dead skin

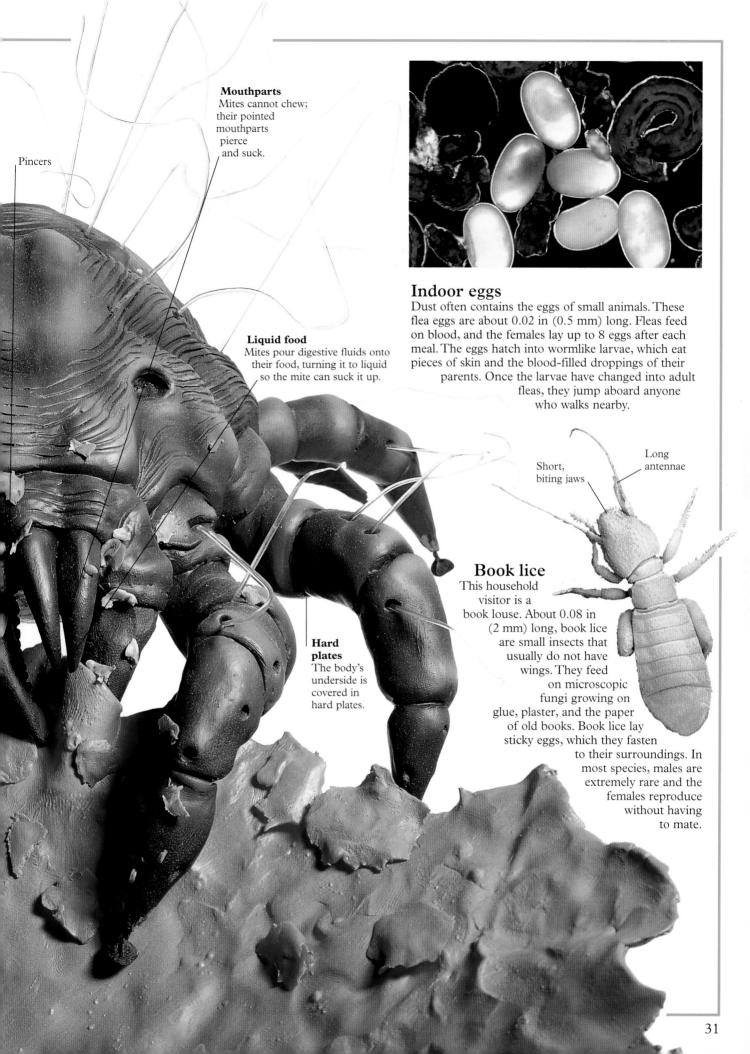

Mouthparts
Mites cannot chew; their pointed mouthparts pierce and suck.

Pincers

Liquid food
Mites pour digestive fluids onto their food, turning it to liquid so the mite can suck it up.

Hard plates
The body's underside is covered in hard plates.

Indoor eggs
Dust often contains the eggs of small animals. These flea eggs are about 0.02 in (0.5 mm) long. Fleas feed on blood, and the females lay up to 8 eggs after each meal. The eggs hatch into wormlike larvae, which eat pieces of skin and the blood-filled droppings of their parents. Once the larvae have changed into adult fleas, they jump aboard anyone who walks nearby.

Short, biting jaws

Long antennae

Book lice
This household visitor is a book louse. About 0.08 in (2 mm) long, book lice are small insects that usually do not have wings. They feed on microscopic fungi growing on glue, plaster, and the paper of old books. Book lice lay sticky eggs, which they fasten to their surroundings. In most species, males are extremely rare and the females reproduce without having to mate.

31

Life on the skin

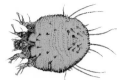

Magnification x 40

Size
Itch mites are about 0.02 in (0.5 mm) long.

For some tiny animals, the surface of a human body is a giant and inviting landscape. It offers warmth, moisture, and an almost endless supply of food. A few of these animals – such as the bed bug – are temporary visitors. After a meal of blood, they crawl or fly away. Others, such as fleas and lice, are permanent residents on the skin. The itch mite has a different way of life. Instead of staying on the surface, it digs a burrow through the skin and lays its eggs inside.

Fine hair
Compared to the itch mite, a hair looks like an immense tree trunk towering overhead.

Cross-section of human skin

Itch mite

Piercing mouth
The mouthparts come together in a sharp point that pierces the surrounding cells.

Burrowing beneath the skin
Itch mites are just big enough to be visible to the naked eye. Male mites spend most of their lives on the surface of the skin and do little harm. However, females excavate winding burrows inside the skin, in which they lay their eggs. This burrowing produces a disease called scabies, which makes the skin feel tender and very itchy. Scabies spreads quickly from person to person.

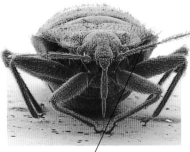

Night visitors
Bed bugs are round, flat-bodied insects about 0.2 in (5 mm) long. During the day thay hide in bed crevices, but at night they emerge when they sense the warmth of a sleeping person. Once a bug has found an area of bare skin, it pierces it with its sharp mouthparts and drinks blood – up to six times its body weight.

Smelly defense
If a bed bug is disturbed, it activates special scent glands and releases a disgusting smell.

Suckers
Related to spiders, itch mites have eight stumpy legs. The front legs end in suckers.

Spines
Backward-pointing spines make it hard to dislodge the female from her burrow.

Living on hair

Lice are tiny blood-sucking insects that are specially adapted for life in human hair, animal fur, or feathers. Human head lice are 0.12 in (3 mm) long and can move rapidly through hair. If disturbed, they hold on to hairs with their claws, and their grip is so strong that they are very difficult to dislodge. Female lice fasten their eggs to hairs with a special gluelike secretion.

Bloodsuckers
The head louse has piercing and sucking mouthparts that withdraw into its head once it has fed.

Hair follicle mite
This mite, enlarged about 650 times, has short legs and a long body.

Fastened egg
Just before hatching, the features of the young louse can be seen through the translucent egg. It emerges through the lid of the egg.

Skin on the hand
An itch mite usually burrows into skin on the wrists and the legs. As it digs, it releases substances into the surrounding skin that cause a rash to spread over the body.

Follicle mites

Microscopic follicle mites are extremely common, living most of the time head downward at the base of hairs, such as eyelashes. At night they sometimes move across the surface of the skin. Unlike itch mites, they do not burrow into the skin and cause no ill-effects. They feed on skin debris and help keep hairs clean.

Beneath the surface
The itch mite burrows just beneath the outermost layer of the skin – made up of dead cells.

Trailing threads
The back legs have hairlike threads on the ends.

Finished burrow
The completed burrow can be up to 0.75 in (2 cm) long.

Burrow home
The female itch mite usually stays in the burrow until she has laid all her eggs, and then dies.

Quick development
The female mite starts to lay eggs when it is only 16 days old.

Laying eggs
The female mite lays about 2 or 3 eggs each day. Once the young mites hatch, they crawl onto the surface.

Nutritious cells
As the mite digs its way through the skin, it uses its mouthparts to pierce living cells, sucking up the fluid they contain.

Living on bread

Magnification
x 100

Size
A pin mold's spore head is about 0.008 in (0.2 mm) across.

If you leave a slice of bread in a damp place, patches of mold will spread across its surface, turning it blue, black, or gray. Molds are microscopic fungi. They usually live on plant or animal remains, but some also infect living things. Like all fungi, they feed with the help of hyphae (slender threads), which digest the sugars and starches that they touch. Once a mold is established, it starts to produce tiny spores that enable it to spread. Some molds make spores in a water droplet, others make dry spores that float away in the air.

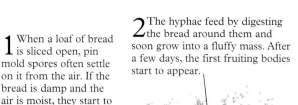

1 When a loaf of bread is sliced open, pin mold spores often settle on it from the air. If the bread is damp and the air is moist, they start to grow slender hyphae.

2 The hyphae feed by digesting the bread around them and soon grow into a fluffy mass. After a few days, the first fruiting bodies start to appear.

Fatal infection
This mold has spread its deadly hyphae through the aphid. It is now ready to release its spores.

Siphuncle (tiny tube) of aphid

Bread mold
Several kinds of mold feed on bread. The one shown here is called pin mold. Its spores form at the tips of fruiting bodies, which look like miniature pins. When the spores are ripe, the fruiting bodies split open and the spores float away.

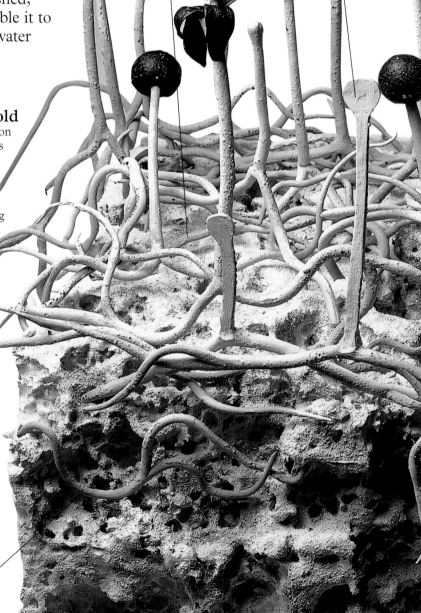

Attacking animals
For some animals – particularly insects and fish – molds can be deadly enemies. This mold is growing out of a siphuncle (tiny tube) on the body of an aphid. The hyphae feed on the substances in living cells. They spread quickly throughout the aphid's body and eventually kill it. Farmers sometimes use molds to kill pests.

Hyphae
The mold's hyphae produce digestive chemicals called enzymes, which break down the starch in the bread.

3 As each fruiting body grows upward, its tip begins to swell. It turns dark as its spores begin to develop.

4 About two days after the fruiting body forms, its spores are almost ripe. The thin membrane around the spores is stretched to the breaking point.

Attacking plants
Here a ripe fungus is breaking out of a breathing hole (stoma) in a dock leaf. Fungi are major killers of plants. They enter plants through wounds in stems and roots, or through the tiny stomata in leaves.

Stoma
The stomata are the weak points in a plant's defenses.

5 Eventually the fruiting body's outer membrane splits. If any of the spores land on another piece of bread, they will produce another patch of mold.

Breaking away
When a new yeast cell breaks away, it leaves a round scar on its parent.

Rapid growth
In good conditions, new cells can bud off every two hours.

Single cell
Yeasts are microscopic fungi that consist of a single cell. They grow by budding off smaller cells, which eventually break away. Yeasts live on substances that contain sugar or starch. They are used to make bread rise and to brew beer.

Useful molds
Not all molds are a nuisance; this mold helps give blue cheese its distinctive flavor. Other molds make substances called antibiotics, which kill bacteria. Antibiotics evolved naturally to give molds a better chance when competing with bacteria for food. They are now important medicines.

Secrets of the soil

Soil makes up one of the richest habitats on Earth. Some of its inhabitants are big enough to see, but most are so small that few people realize that they are there. These tiny forms of life include bacteria and fungi, which live by breaking down dead remains. They also include many kinds of animals, from harmless vegetarians to stealthy hunters that pounce on their victims in complete darkness. Pseudoscorpions are some of the most common of these miniature predators. Unlike true scorpions, they have poisonous claws instead of stinging tails. Guided by sensitive hairs, they slowly clamber toward smaller animals, and then grab them with their claws. Toothlike projections at the tips of their claws inject powerful poison, giving the unlucky victim little chance of escape.

Magnification x 9

Size
Most pseudoscorpions are less than 0.1 in (3 mm) long.

Predatory fungi
Most soil fungi live on decaying remains, but some have another way of feeding – they grow microscopic nooses that can trap tiny worms. This roundworm has accidentally tried to wriggle through a noose. The noose has tightened, trapping the worm.

Jointed legs
Pseudoscorpions have four pairs of jointed legs.

Hidden spring
The springing organ is attached to the underside of the springtail's abdomen.

Slow progress
Springtails normally move slowly on their stubby legs.

Chewing food
A springtail uses its biting mouthparts to chew up plant remains.

Springtail
Primitive insects called springtails abound in the soil, where they are often attacked by pseudoscorpions. If a springtail is cornered underground, it has little chance of escape. However, on the surface it can catapult itself to safety by flicking a special spring-loaded tail, which is normally tucked underneath its body.

Hitching a ride
This pseudoscorpion is using its claws to hitch a ride on a fly – a useful way for the slow-moving pseudoscorpion to get around. Each claw is strong enough to support its own weight. This hitchhiker is not as harmless as it might look, because its poison sometimes kills the animal giving it a ride.

Body segments
The body is covered by hard plates that hinge together.

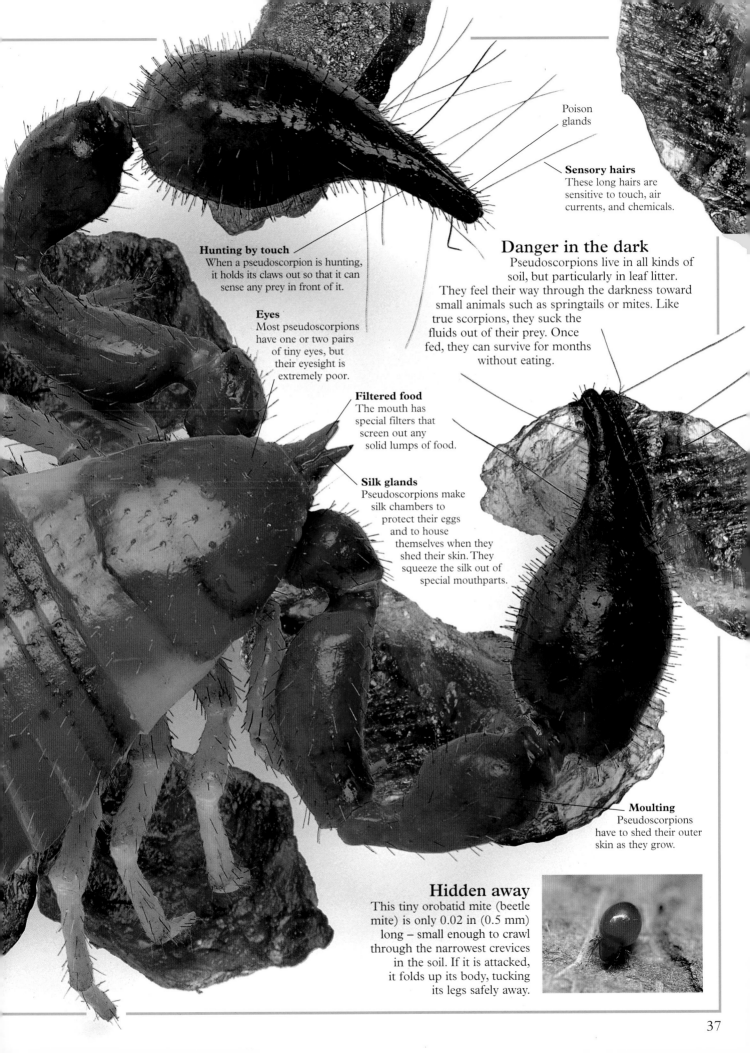

Poison
glands

Sensory hairs
These long hairs are
sensitive to touch, air
currents, and chemicals.

Danger in the dark
Pseudoscorpions live in all kinds of
soil, but particularly in leaf litter.
They feel their way through the darkness toward
small animals such as springtails or mites. Like
true scorpions, they suck the
fluids out of their prey. Once
fed, they can survive for months
without eating.

Hunting by touch
When a pseudoscorpion is hunting,
it holds its claws out so that it can
sense any prey in front of it.

Eyes
Most pseudoscorpions
have one or two pairs
of tiny eyes, but
their eyesight is
extremely poor.

Filtered food
The mouth has
special filters that
screen out any
solid lumps of food.

Silk glands
Pseudoscorpions make
silk chambers to
protect their eggs
and to house
themselves when they
shed their skin. They
squeeze the silk out of
special mouthparts.

Moulting
Pseudoscorpions
have to shed their outer
skin as they grow.

Hidden away
This tiny orobatid mite (beetle
mite) is only 0.02 in (0.5 mm)
long – small enough to crawl
through the narrowest crevices
in the soil. If it is attacked,
it folds up its body, tucking
its legs safely away.

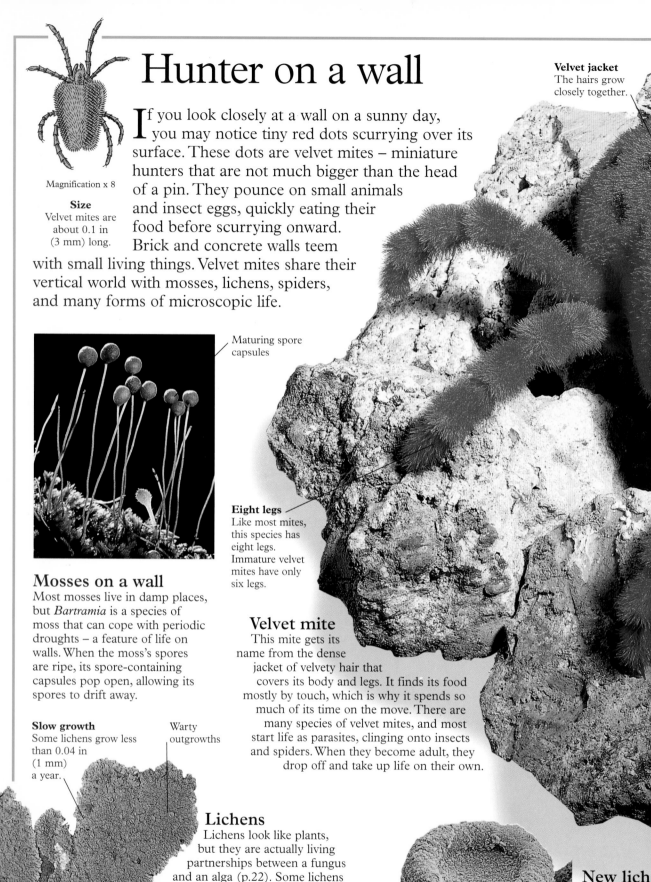

Hunter on a wall

Magnification x 8

Size
Velvet mites are about 0.1 in (3 mm) long.

If you look closely at a wall on a sunny day, you may notice tiny red dots scurrying over its surface. These dots are velvet mites – miniature hunters that are not much bigger than the head of a pin. They pounce on small animals and insect eggs, quickly eating their food before scurrying onward. Brick and concrete walls teem with small living things. Velvet mites share their vertical world with mosses, lichens, spiders, and many forms of microscopic life.

Velvet jacket
The hairs grow closely together.

Maturing spore capsules

Eight legs
Like most mites, this species has eight legs. Immature velvet mites have only six legs.

Mosses on a wall
Most mosses live in damp places, but *Bartramia* is a species of moss that can cope with periodic droughts – a feature of life on walls. When the moss's spores are ripe, its spore-containing capsules pop open, allowing its spores to drift away.

Velvet mite
This mite gets its name from the dense jacket of velvety hair that covers its body and legs. It finds its food mostly by touch, which is why it spends so much of its time on the move. There are many species of velvet mites, and most start life as parasites, clinging onto insects and spiders. When they become adult, they drop off and take up life on their own.

Slow growth
Some lichens grow less than 0.04 in (1 mm) a year.

Warty outgrowths

Lichens
Lichens look like plants, but they are actually living partnerships between a fungus and an alga (p.22). Some lichens are bright orange; others are green, black, or gray. Lichens grow very slowly but are very hardy. They do not need soil and can extract all the minerals they need from bare walls and rocks.

New lichen
This magnified fruiting body spreads a lichen's spores. Lichens also reproduce when warty outgrowths break off to form new lichens.

Egg-laying
Female velvet mites scatter their eggs on walls and soil.

Pinned down
The springtail is held down so it cannot escape.

Polar predator
Dependent on touch rather than sight, this Antarctic mite attacks rapidly when it bumps into its prey. It grasps a springtail with its front legs and pierces it with its needlesharp mouthparts. The predator sucks fluid from the springtail's body. Since this meal can last several hours, the mite carries the springtail around until it looks like a deflated bag, drained of all liquid.

Mouthparts
Pincerlike mouthparts immobilize the velvet mite's prey.

Zebra spiders' eyes
A zebra spider has two main eyes, and several smaller ones on either side.

Giant jumps
Despite having lots of eyes, most spiders are not very good at seeing animals around them. However, a zebra spider has two extra-large eyes that point forward like a car's headlights. The spider uses them to spot small animals, on which it leaps with the agility of an acrobat. About 0.2 in (5 mm) long, the zebra spider can be seen jumping on sunlit walls and wooden fences.

Feeling the way
The mite uses its front legs to feel for food and often holds them off the ground.

Hidden food
Insects often lay their eggs in crevices in walls, where the velvet mite can find them.

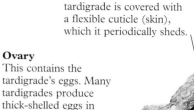

Supreme survivors

Size
A typical tardigrade
(water bear) is
about 0.015 in
(0.4 mm) long.

Imagine living in a bath of steaming acid or a pool of scalding mud. Imagine surviving without water for over 20 years, or withstanding temperatures lower than −148°F (−100°C). For humans, all these things are impossible, but for some forms of microscopic life – such as bacteria, tardigrades, and rotifers – they present few problems. Some are adapted for life in the world's most hostile environments, while others survive in places where conditions can suddenly change.

Cuticle
This smooth-bodied tardigrade is covered with a flexible cuticle (skin), which it periodically sheds.

Ovary
This contains the tardigrade's eggs. Many tardigrades produce thick-shelled eggs in harsh conditions.

Extreme habitats
Volcanic mud pools and hot sulfur-filled springs are dangerous environments for most living things. However, they are filled with dissolved chemicals on which heat-loving bacteria thrive. These bacteria can only live in temperatures above 131°F (55°C).

Colored by food
Some tardigrades get their colors from the food they eat.

Internal fluid
Tardigrades do not have blood, but their body spaces are filled with fluid.

Solitary confinement
Each cyst contains a single amoeba. The cyst breaks open when conditions improve.

Hard times ahead
Many living things have a stage in their life cycle that can survive difficult conditions. For insects, this stage is usually the egg, but for amoebas, it is a structure called a cyst. Cysts can remain dormant for over a decade.

Cryptobiosis
Many tiny tardigrades live in the thin film of moisture that covers moss. If this film starts to dry up, a tardigrade enters a state called cryptobiosis, meaning "hidden life." Its body slowly dehydrates, and almost all of its chemical processes come to a halt. It can survive in this state for many decades.

One of eight stubby legs

Cysts have hard, waterproof walls

Exoskeleton
A covering stops the mite from drying out.

The big freeze
This mite lives in the Antarctic Peninsula, where winter temperatures sometimes drop to −22°F (−30°C). It becomes dormant (inactive) when cold weather sets in. Its body contains chemicals that work like antifreeze and prevent dangerous ice crystals from forming inside it.

Claws
Each foot ends in a collection of clawlike hooks.

Claw gripping a moss stem

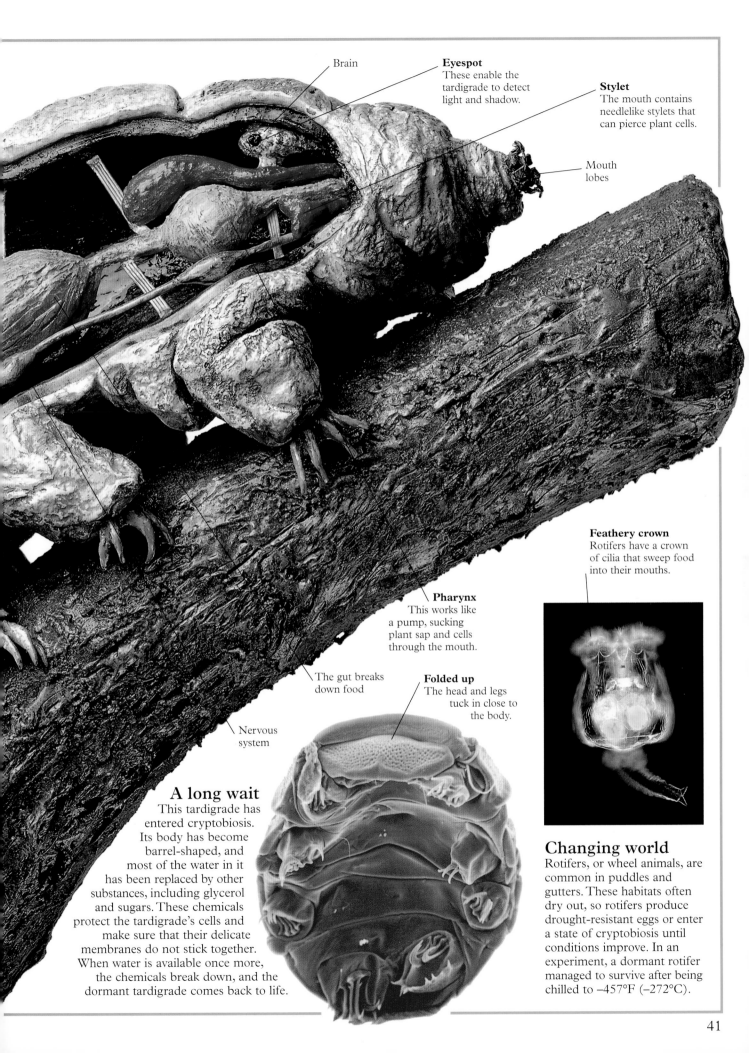

Brain

Eyespot
These enable the
tardigrade to detect
light and shadow.

Stylet
The mouth contains
needlelike stylets that
can pierce plant cells.

Mouth
lobes

Pharynx
This works like
a pump, sucking
plant sap and cells
through the mouth.

The gut breaks
down food

Folded up
The head and legs
tuck in close to
the body.

Nervous
system

Feathery crown
Rotifers have a crown
of cilia that sweep food
into their mouths.

A long wait
This tardigrade has
entered cryptobiosis.
Its body has become
barrel-shaped, and
most of the water in it
has been replaced by other
substances, including glycerol
and sugars. These chemicals
protect the tardigrade's cells and
make sure that their delicate
membranes do not stick together.
When water is available once more,
the chemicals break down, and the
dormant tardigrade comes back to life.

Changing world
Rotifers, or wheel animals, are
common in puddles and
gutters. These habitats often
dry out, so rotifers produce
drought-resistant eggs or enter
a state of cryptobiosis until
conditions improve. In an
experiment, a dormant rotifer
managed to survive after being
chilled to −457°F (−272°C).

Glossary

Red velvet mite

A

Amoeba
A common single-celled organism with an irregular shape. Amoebas thrive in pond and soil water and move by changing shape.

Aphid
A small insect that lives by sucking sap from plants.

B

Bacterium
A very small living thing with a single, simple cell. Disease-causing bacteria are often known as germs.

C

Cell
The smallest complete unit of living matter. All cells are surrounded by thin membranes, and some also have tough cell walls.

Cellulose
A special sugar that plants use to build their cell walls. Unlike most sugars, it forms tough strands.

Cilia on *Vorticella*

Chloroplast
A special structure in plant cells that traps the energy in sunlight.

Chromosome
A package of tightly folded DNA inside a living cell.

Cilium
A microscopic hairlike projection from a cell. Some microbes have thousands of cilia. These beat in sequence to move the microbe backward and forward.

Crustacean
An animal with a hard body case and two pairs of antennae (feelers). Most crustaceans live in water.

Cryptobiosis
A special state that some animals use to survive difficult times. During cryptobiosis, most of their body processes come to a halt.

Cyst
A tough resting stage in the life cycle of many microorganisms.

Cytoplasm
The living matter inside a cell. It consists of a jellylike fluid and often a variety of organelles.

D

Dehydration
Severe water loss.

DNA (Deoxyribonucleic acid)
A chemical that holds the instructions needed to build living things and to make them work.

Dormant
Completely inactive. Many living things – including animals and plants – become dormant when times are hard.

E

Enzyme
A substance that speeds up a vital chemical reaction in a living organism.

Eyespot
A simple sense organ that can detect light and shade.

F

Fertilization
The joining together of a male and a female cell to produce a new living thing.

Filament
A long, thin strand.

Flagellum
A special hairlike organ that beats from side to side to move a cell along. A flagellum is often as long as the cell.

Follicle
In skin, a small hollow that contains the root of a hair.

Fruiting body
A special reproductive structure that forms spores and helps them spread. Some fruiting bodies are easy to see, but others are microscopic.

Virus

Fungus
A form of life that grows a network of slender threads that absorb food from their surroundings.

H

Hypha
A slender feeding thread produced by a fungus.

L

Larva
A young animal that develops by changing shape. Larvae usually look very different from their parents.

M

Membrane
A thin, flexible layer. Living cells are surrounded by membranes, and so are some of the organelles inside them.

Microbe
Often used to describe any living thing that cannot be seen with the naked eye. Also known as microorganisms.

Micrograph
A picture taken with a microscope.

Pin mold on a slice of bread

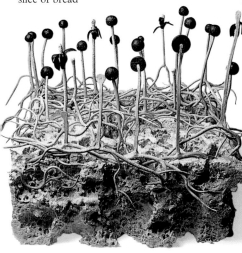

Microorganism
The scientific term for any organism that is too small to be seen with the naked eye.

Mite
A small animal, related to spiders, that lives on liquid food from living things or dead remains.

Mold
A kind of fungus that often forms furlike patches when it grows.

N

Nucleus
The control center of a cell. A nucleus contains the cell's chemical instructions, stored in DNA. It is separated from the cytoplasm by a nuclear membrane.

O

Organelle
A structure inside a cell that carries out a particular task.

Organism
Any living thing.

P

Plankton
The name used for tiny living things that float near the surface of water. Plankton includes animals, plants, and algae.

Protein
A complex chemical made by living things. There are many kinds of proteins and they have many uses, from building cells to speeding up chemical reactions.

Protozoan
A microscopic living thing that has a single cell with a nucleus and lives by taking in food.

Pseudopod
A footlike outgrowth of a cell.

R

Ribosome
A structure inside a cell that is used to make proteins.

S

Seed
A tough structure used by many plants to reproduce. Each one contains a plant in its early stages and a food store, and is surrounded by a tough coat.

White blood cell attacking bacteria

Silica
A mineral substance used by many microorganisms to make shells and skeletons.

Siphuncle
In aphids, a short tube

Protozoan

projecting from the body. It produces chemicals to fend off attackers.

Species
A collection of living things that look alike and behave in a similar way. The members of a species can breed together in the wild. Each one has a two-part scientific name that is usually printed in *italic* type.

Spore
A microscopic seedlike object that is used in reproduction. Unlike a seed, a spore often contains just one cell.

Stigma
The female part of a flower that collects pollen grains.

Stoma
A microscopic hole that allows gases to flow in and out of a leaf.

Stylet
A piercing part of an animal's mouth.

V

Virus
A package of chemicals that can infect a living cell and make the cell produce copies of the virus.

Index

Acknowledgments

The publisher would like to thank:
Dr. Bloch, British Antarctic Survey; Dr.
Brain and John Pacy, King's College,
London; Dr. Burns, Leicester Royal
Infirmary; Geoff Hancock, the Glasgow
Museums; Janet Hurst, Society of
General Microbiology; Dr. Mark
Judson, Muséum National d'Histoire
Naturelle, Paris; Steve Moss, British
Mycological Society, and Alec Shaw
Stewart for their time and advice.
Nancy Jones, Shirin Patel, Sean
Stancioff for editorial assistance. Dave
Morgan for photographic assistance.

Additional photography: Andy
Crawford, Dave King

Additional models: Wendy Chander

Photoshop retouching: Bob Warner

Visualization: Jill Plank

Artwork: John Woodcock

Picture Credits:
Key: t=top; b=below; c= center; l=left;
r=right
Biophoto Associates: 12cl, 38cl.
British Antarctic Survey: 39tr, 40bl.
EM Unit, King's College, London:
Dr. Tony Brain 16bl, 16cr, 18bcl, 27tl,
27tr, 32bl; John Pacy 18b, 31br, 34bl.
Leicester Royal Infirmary: Dr. D.A.

Burns 33tl, 33tc. **Microscopix
Picture Library:** Andrew Syred 8tra,
8trb, 12bl. **Natural History Museum,
London:** 36–7. **Oxford Scientific
Films:** 21cr; London Scientific Films
36tr, 39br; Alastair MacEwen 31tr;
Peter Parks 16bc, 36bl; Kim Westerskov
40cl. **Robert O. Schuster courtesy
of Dr. R. Diane Nelson** 41bc. **Planet
Earth Pictures:** Pete Atkinson 22cl;
A.W. Raksoy Blackstar 9tl, 22cr; Steve
Hopkins 37br. **Science Photo
Library:** /Biophoto Associates 10br,
29cr; Dr. Jeremy Burgess 14bl, 16tr,
28cl, 35cl, 38br, 38bl; Chris Bjornberg
15tr; C.N.R.I 14bcr, 35br; Barry
Dowset 19bcr; Eye of Science 9tr, 14cl;

Eric Grave 24bl; Manfred Kage 22tr,
24cl, 25bl, 35cr, endpapers; Dr. Kari
Lounatmag 9tc; Dr. Gopal Murti 10bcl;
Alfred Pasieka 13tr, 16acr; Dr. David
Patterson 40 bcl; Professor A. Polliak
21tl; J.C. Revy 29br; David Scharf 28bl;
Microfield Scientific 35tr; Andrew
Syred 23br, 30cl, 33tr; M.I. Walker
26bl; John Walsh 41bcr; L. Willatt 11br.

Every effort has been made to trace the
copyright holders. DK apologizes for
any unintentional omissions and would
be pleased, in such cases, to add an
acknowledgment in future editions.

Index: Hilary Bird

DATE DUE

DEMCO, INC. 38-3011

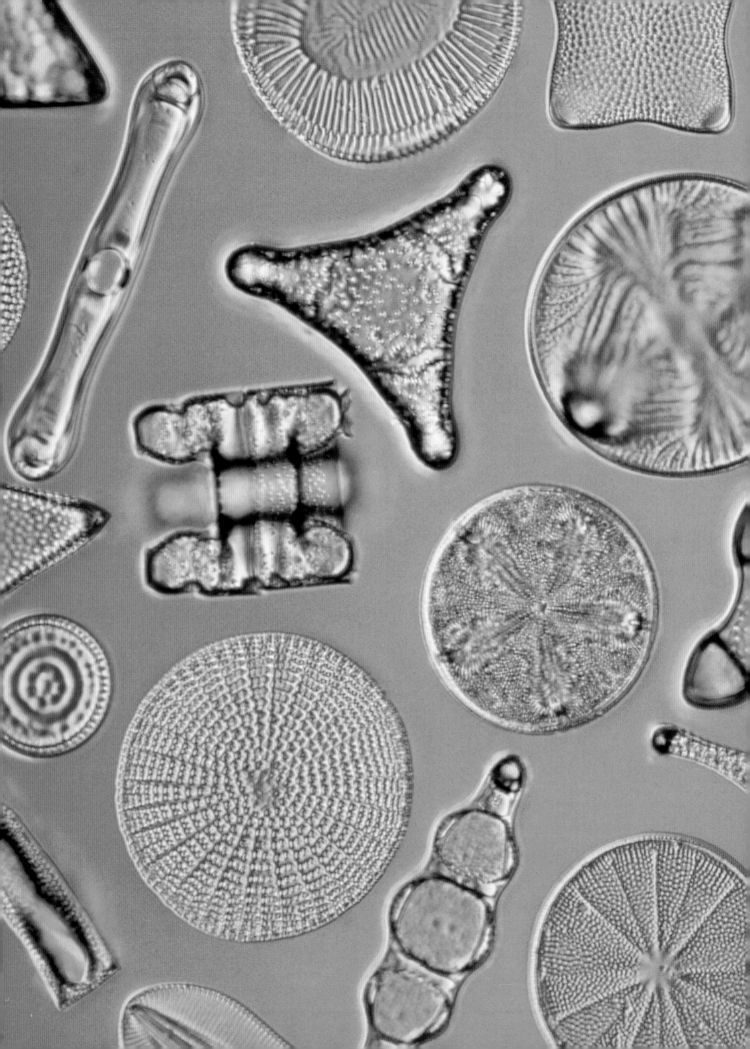